Engineering and the Mind's Eye

Engineering and the Mind's Eye

Eugene S. Ferguson

The MIT Press
Cambridge, Massachusetts
London, England

Set in Times by Achorn Graphic Services.
Printed and bound in the United States of America.

Library of Congress Cataloging-in-Publication Data
Ferguson, Eugene S.
Engineering and the mind's eye / Eugene S. Ferguson.
p. cm.
Includes bibliographical references and index.
ISBN 0-262-06147-3
1. Engineering. 2. Creative thinking. I. Title.
TA145.F37 1992   91-42833
620—dc20   CIP

This one, at long last
and with abiding love,
is for Jo.

Contents

I know very well that in a great many circles the man who does not enter with a neatly arranged plan, with a set of doctrines, with a rounded and sonorous formula, and with assurance about everything, is set down as something of an old fogey, perhaps reactionary, certainly not one of the elect who are "doing things" and providing guidance for the race. I must assume the risk. I have no formula. [But I shall resist] those who have the formula for so many things and who seek so avidly to force it down the throats of every one else.

Matthew Woll, "Standardization," *Annals* (American Academy of Political and Social Science, Philadelphia) 137 (May 1928), p. 47

Preface

This scientific age too readily assumes that whatever knowledge may be incorporated in the artifacts of technology must be derived from science. This assumption is a bit of modern folklore that ignores the many nonscientific decisions, both large and small, made by technologists as they design the world we inhabit. Many objects of daily use have clearly been influenced by science, but their form, dimensions, and appearance were determined by technologists—artisans, engineers, and inventors—using nonscientific modes of thought. Carving knives, bridges, clocks, and aircraft are as they are because over the years their designers and makers have established their shapes, styles, and textures.

Many features and qualities of the objects that a technologist thinks about cannot be reduced to unambiguous verbal descriptions; therefore, they are dealt with in the mind by a visual, nonverbal process. The mind's eye is a well-developed organ that not only reviews the contents of a visual memory but also forms such new or modified images as the mind's thoughts require. As one thinks about a machine, reasoning through successive steps in a dynamic process, one can turn it over in one's mind. The engineering designer, who brings elements together in new combinations, is able to assemble and manipulate in his or her mind devices that as yet do not exist.

If we are to understand the nature of engineering, we must appreciate this important although unnoticed mode of thought. It has been nonverbal thinking, by and large, that has fixed the outlines and filled in the details of our material surroundings. In their innumerable choices and decisions, technologists have determined the kind of world we live in, in a physical sense. Pyramids, cathedrals, and rockets exist not because of geometry, theory of structures, or thermodynamics, but because they were first pictures—literally visions—in the minds of those who conceived them.

This book attempts to clarify the nature and significance of nonverbal thought in engineering. It argues that modern engineering—that is, the engineering of the last 500 years—has depended heavily and continuously on nonverbal learning and nonverbal understanding. Until the second half of the twentieth century, engineering schools taught an understanding of engineering drawings by teaching how to make such drawings; they built an appreciation of the nature of materials and machines through laboratory experience. They understood that most of an engineer's deep understanding is by nature nonverbal, the kind of intuitive knowledge that experts accumulate.

Since World War II, the dominant trend in engineering has been away from knowledge that cannot be expressed as mathematical relationships. The art of engineering has been pushed aside in favor of the analytical "engineering sciences," which are higher in status and easier to teach. The underlying argument of this book is that an engineering education that ignores its rich heritage of nonverbal learning will produce graduates who are dangerously ignorant of the myriad subtle ways in which the real world differs from the mathematical world their professors teach them.

Publisher's Notes

Some of the figures are grouped or paired thematically. It is hoped that the reader will put up with the occasional inconvenience of having to flip a few pages in the interest of having related figures presented in close proximity to one another.

Citations prefixed with the letter f (e.g., [f3]) correspond to the "Notes to Figures," which begin on page 229.

Preface

Acknowledgments

In any direction my mind's eye looks, I see people who have helped me with this book. Scholars who for 30 years have been discovering and writing about the nature of technology and engineering have extended my horizons and answered many interesting questions I didn't know how to ask. The reader will meet many of those scholars in the text and the notes of this book. They have been important to me; I hope others will want to learn more about those they had not encountered.

The gratifying response to my article in *Science* ("The Mind's Eye: Nonverbal Thought in Technology," August 26, 1977) convinced me that I was not the only person interested in nonverbal thinking. I thank those who took the time and trouble to write. That article turned out as it did because of the generous help of my colleagues and friends George Basalla, Barbara Benson, Lynwood Bryant, and David Hounshell. The justifiably skeptical response of the late Bertha Leaman to an earlier talk on Renaissance engineering was crucial to the development of this book. She encouraged me to rethink the significance of technical drawings in that period, a process which eventuated in the *Science* article and, lo, these many years later, this book. Although I failed to acknowledge his influence on the article, Edwin Layton was (in retrospect) the individual who persuaded me that engineering is founded on nonverbal thought. Merritt Roe Smith suggested the book, and Larry Cohen has waited for it years longer than he expected to.

Svante Lindqvist contributed good advice, not all of it followed because I didn't know how. Gustina Scaglia gave freely of her broad and deep knowledge of Renaissance engineers. James W. Althouse spent many hours answering my questions and raising others about engineering design. Elwin C. Robison let me quote a letter he wrote me on graphic statics. E. Hubbard Yonkers encouraged my work and let me have one of his "talking" sketches. Henry Petroski has given me constant, generous encouragement and good advice; he has also saved me from some technical blunders. For 20 years, the strengths of the University of Delaware Library have been, for me, embodied in Nathaniel Puffer, who enlarged my precincts as he built solid collections of rare books and historic serials in engineering and technology. Charles M. Haines has been constantly available to explain to me ideas (often his) that I was trying to articulate. Stuart W. Leslie gave me timely criticism that finally set straight in my mind what the book is about. "Book doctor" Catherine E. Hutchins patiently helped me turn an ailing manuscript into a much healthier one. I cannot even *imagine* a better editor than Paul M. Bethge. George Basalla is the person without whose encouragement and help at every turn this book would not have been completed. Many other colleagues, correspondents, and friends have put their marks on this book, but I am either too forgetful or insufficiently perceptive to recall or recognize their influence. To those: please accept my apologies and thanks.

Finally, I have been fortunate beyond any reasonable expectations to have had a series of employers who have been perfect patrons in that they found the means and the tolerance to support me and my projects: the late Henry M. Black, Robert P. Multhauf, the late George R. Town, Walter J. Heacock, John A. Munroe, Willard A. Fletcher, Stephen Salsbury, and Richard L. Bushman. Marjorie Johnson Tilghman, teacher and friend, has encouraged and helped me in many ways over many years. Jo Mobley Ferguson, my wife of 43 years, has held the enterprise together.

Engineering and the Mind's Eye

1 The Nature of Engineering Design

From the point of view of modern science, design is nothing, but from the point of view of engineering, design is everything. It represents the purposive adaptation of means to reach a preconceived end, the very essence of engineering.

Edwin T. Layton, Jr., 1976[1]

It is usually a shock to [engineering] students to discover what a small percentage of decisions made by a designer are made on the basis of the kind of calculation he has spent so much time learning in school.

"Report on Engineering Design," 1961[2]

One of the earliest and best definitions of engineering, in the 1828 charter of the (British) Institution of Civil Engineers, asserts that engineering is "the art of directing the great sources of power in nature for the use and convenience of man."[3] That definition is still accurate and adequate. The great sources of power—fire and falling water, augmented slightly in the latter half of the twentieth century by nuclear fission—are being used by engineers today, as they were in 1828, to change the world for the "use and convenience" of humankind.[4]

It is important for all of us to understand how engineers choose and plan the changes they make, because engineers have an effect upon the kind of world we live in out of all proportion to their numbers. In America and Western Europe engineers constitute less than 1 percent of the population, yet because engineers are the ones who design the bridges, highways, automobiles, airplanes, telephone systems, water systems, heating and air conditioning systems, computers, and television networks—things that influence strongly and directly the way we live from day to day—they are far more influential than their numbers suggest.

Many engineers deny their influence, insisting that they merely carry out the orders of others—politicians, for instance. Yet in fact it is the engineers who draw up the politicians' shopping lists by furnishing specific solutions to particular problems, complete with plans and specifications. And of course the solutions proposed by engineers require engineers to carry them out. For example, three years before 1961, when President John Kennedy called for the United States to put a man on the moon to outdo the Russians, engineers at NASA had proposed such an expedition as the ninth step in a comprehensive space program.

The crucial problem for the engineers was not to determine how to get to the moon but rather to find a patron who would furnish the money and give the program political legitimacy. In April 1961, when the Russian Yuri Gagarin became the first astronaut to go around the world in orbit, Kennedy's advisers took some of the NASA proposals off the shelf in order to meet the president's urgent need to "catch up" with the Russians in space and to restore confidence in US technological superiority. On May 25, Kennedy announced an ambitious space program that included a manned expedition to the moon.[5]

A few years earlier, in the mid 1950s, a series of television shows produced by Walt Disney had promoted the ideas of Wernher von Braun and other space enthusiasts. In March of 1955, President Dwight Eisenhower telephoned Disney to borrow copies of one program, entitled "Man in Space," to run for important audiences in the Pentagon. On July 30, 1955, Eisenhower announced plans to launch a satellite during the International Geophysical Year, 1957–58. According to David R. Smith, director of the archives at Walt Disney Publications, "Wernher von Braun never forgot the boost [the Disney shows] had given his efforts. On the day in 1968 when Apollo first circled the moon, he placed a telephone call to Ward Kimball [producer of the shows], saying, 'Well, Wahd, it looks like they're following our script.'"[6]

In order to produce a new machine, structure, or other technological artifact, two separate but closely related processes are generally required. In the first, engineering designers convert the visions in their minds to drawings and specifications. In so doing, they solve an ill-defined problem that has no single "right" answer but has many better or worse solutions. Engineers learn a great deal during the process of design as they strive to clarify the visions in their minds and seek ways to bring indistinct elements into focus. When the designers think they understand the problem, they make tentative layouts and drawings, analyze their tentative designs for adequacy of performance, strength, and safety, and then complete a set of drawings and specifications. The second process revolves around the finished drawings and specifications. Those who will make or build the machine, structure, or system can now learn exactly what they are expected to produce. Until their task is complete and the project has been turned over to its user, those drawings and specifications will be the formal instructions that guide their work.

Engineering drawings are expressed in a graphic language, the grammar and syntax of which are learned through use; it also has idioms that only initiates will recognize. And because the drawings are neatly made and produced on large sheets of paper, they exude an air of great authority and definitive completeness.

Although the drawings appear to be exact and unequivocal, their precision conceals many informal choices, inarticulate judgments, acts of intuition, and assumptions about the way the world works. The conversion of an idea to an artifact, which engages both the designer and the maker, is a complex and subtle process that will always be far closer to art than to science.

Designing without Drawings: The Artisan's Way
In order to ponder what must be in any design, whether it is drawn only in an artisan's mind and worked out directly in suitable materials or whether it is composed on a computer screen and reduced automatically to a working drawing, it will be useful to start by looking at how artisans work.

Before a thing is made, it exists as an idea. The idea may be a clear vision or it may be little more than a glimpse of a possibility. If the idea is in the head of an artisan, he can make the thing directly, requiring only the materials, tools, and skill needed to convert the materials to the desired thing. The distinctive American axe of the eighteenth and nineteenth centu-

4

ries originated when a blacksmith modified the design of traditional European trade axes (figures 1.1–1.3). The head was made heavier all over, and increasing the proportion of metal in the poll radically improved the balance of the axe for heavy chopping and for felling trees. The new style was further modified by other blacksmiths who heeded the suggestions and criticisms of experienced axemen.[7]

An artisan might make a sketch of an idea on paper or on the material of which the thing is to be made, in order to keep in mind a shape or configuration of components, or might move immediately to make a model of the idea, as some boatbuilders do. In any case, the chosen materials have an active role in design, modifying an artisan's vision as he finds he misjudged some aspect of the qualities of the materials. Finally, as in the case of the axe, the user of the thing may also have an active role to play—sometimes while the design is being worked out, but more often afterward in reports on the performance of the finished thing.

The new owner of a small fishing craft may return to the builder's shop and comment: "Joe, that damn boat almost killed me crossing the bar today. You've got to cut away the forefoot a little and put some beam [width] into her, back aft."[8] The builder may or may not modify that particular boat, but almost certainly the next boat will be different. The builder-designers, the owners, and the local geography all contribute to the evolution of a "type" of fishing boat.

Designing with Drawings: The Engineer's Way
If the idea for the thing to be made is not in an artisan's head but in an engineer's, the engineer must have some means of explaining that vision to the worker who will construct it. For more than 500 years, engineers have made increasing use of drawings to convey to workers what is in their heads. Some of the earliest drawings of machines to be built by workers not under the designer's immediate supervision were made by Filippo Brunelleschi, who in the early fifteenth century designed and supervised the building of the great masonry dome of the cathedral of Santa Maria del Fiore in Florence. Wishing to keep his original design for a crane from being pirated,

Brunelleschi sent drawings of various components to separate shops outside the city and assembled the parts when he received them.[9]

The differences between the direct design of the artisan and the design drawing of the engineer are differences of format rather than differences of conception. In both cases the design starts with an idea—sometimes distinct, sometimes tentative—which can be thrown on the mind's screen and observed and manipulated by the mind's eye. Usually, the "big," significant, governing decisions regarding an artisan's or an engineer's design have been made before the artisan picks up his tools or the engineer turns to his drawing board. Those big decisions have to be made first so that there will be something to criticize and analyze. Thus, far from starting with elements and putting them together systematically to produce a finished design, both the artisan and the engineer start with visions of the complete machine, structure, or device.

A complex modern device such as an internal-combustion engine is usually designed by a team of engineers, the specialized knowledge of each one contributing to meeting the various requirements of the overall problem. The team is invariably led by an engineer who keeps the overall design constantly in mind, even as unanticipated problems force its modification.

The set of drawings for an engine may number several hundred. Each separate part of the engine is drawn in full detail in order to show its shape, its dimensions, and all its other features (see, for example, figure 1.4). Subassembly drawings show how individual parts are to be arranged and joined; assembly drawings locate subassemblies in a proper overall configuration.

The significant verb in this description of drawings is *show*. The drawings have two principal purposes. First, they show designers how their ideas look on paper. Second, if complete, they show workers all the information needed to produce the object. The information that the drawings convey is overwhelmingly visual: not verbal, except for notes that specify materials or other details; not numerical, except for dimensions of parts and assemblies. Such drawings, resulting from nonverbal thinking and possessing the ability to transfer visual information across space and time, are so constantly

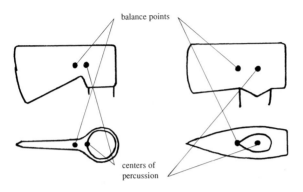

Figure 1.1
Left: European trade axe. Right: American axe.

Figure 1.2
Michigan pattern 4-pound axehead.

Chapter 1

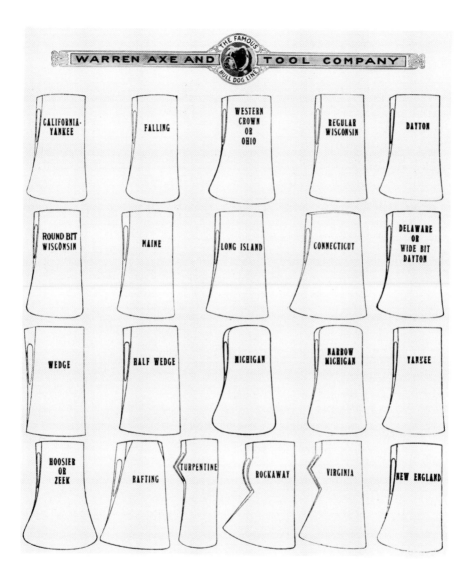

Figure 1.3
Axehead patterns in 1916 trade catalogue.

The Nature of Engineering Design

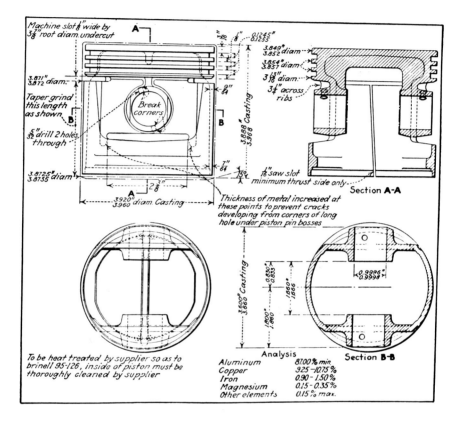

Figure 1.4
Working drawing of aluminum alloy piston for Ford Model A automobile engine,
1928.[fl]

present in offices and shops that their crucial role as intermediaries of engineering thought is easily overlooked.

To build an engine after it has been designed requires the special knowledge and skills of several kinds of workers: patternmakers, foundrymen, machinists, and die makers, to name only those who will work on the block, the crankshaft, the connecting rods, the valves, and similar components. Because the engineering drawings can be duplicated and distributed to many shops, all the parts requiring (for example) machinists' skills need not be made by the same workers or even in the same machine shop. Thus the set of drawings makes possible the organization and timing of fabrication and assembly.

The design of the engine will inevitably have been modified as the engineers wrestled with unanticipated difficulties that appeared only when the "paper parts" were converted to metal ones. Myriad design choices will have been made to orchestrate the operation of the various components of the engine. Some of the choices will have been wrong. Yet making wrong choices is the same kind of game as making right choices; there is often no *a priori* reason to do one thing rather than another, particularly when neither has been done before. No bell rings when the optimum design appears. The principles and techniques of engineering design can never be fully articulated, however much those promoting a "design science" may believe that a designer's judgment can be incorporated in a general-purpose computer program.[10]

Engineering Knowledge

The formal knowledge that engineering designers use is not science, although a substantial part of it is derived from science. It includes as well knowledge based on experimental evidence and on empirical observations of materials and systems. Walter Vincenti, an aeronautical engineer who has traced the evolution of engineering knowledge, argues cogently that it has been developed and formalized primarily to meet the needs of engineering designers.[11] For example, the optimum or "correct" degree of inherent stability of an airplane was by no means obvious until more than 30 years after the Wrights' first powered flight in 1903. European designers of airplanes assumed at first that pilots would merely steer their craft in the manner of automobile drivers or mariners. Therefore, the need for inherent stability seemed obvious. On the other hand, as the Wrights saw, too much inherent stability would reduce a pilot's control of his airplane. A bicycle is inherently unstable, yet with practice it is readily controlled. But, as Vincenti reminds us, training wheels on a bicycle, intended to help hold the bike upright for the beginning rider, are soon discarded as the rider's reflexive responses make them intrusive rather than helpful.

The Wright brothers had recognized that airplanes, unlike automobiles and boats, must be controlled in three dimensions rather than merely steered. Their decision to build airplanes that would require skilled piloting was, in Vincenti's words, "largely deliberate, conceptually linked to the sideways instability of the bicycle, with which the Wrights were familiar." They devised an ingeniously simple and elegant system of wing warping to keep their first airplanes on an even keel and to allow banked turns (figure 1.5).[12]

By the 1930s, designers, pilots, aerodynamicists, and instrumentation specialists had reached a consensus that an aircraft should have enough stability to avoid disaster through a momentary aberration but enough instability to give the pilot optimum control. Vincenti points out that although aerodynamicists (scientists, if you please) were involved in the debates, the subjective response of pilots—a sense of what is flyable—and the experience of designers were the determining factors in the consensus. It was a collective "practical judgment (based largely on subjective opinion) of a sort that cannot be avoided in engineering"—"an instance *par excellence* of engineering, as opposed to scientific, knowledge."[13] Eventually the consensus was codified in reasonably unambiguous terms and made a routine part of design specifications.[14]

The historian Edwin Layton has contributed to the topic of engineering knowledge the important insight that what engineers call "the engineering sciences"—mechanics, thermodynamics, materials science, and several others—have taken their patterns from science. They are mathematical and exact within prescribed limits, and their similarities to the "hard sciences" are so striking that Layton calls science and the engineering sciences "mirror-image twins."[15] The purpose of the engineering sciences, however,

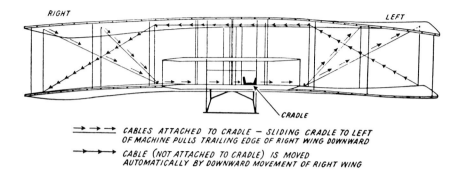

Figure 1.5
Wing warping in the first Wright airplane, 1903. To maintain lateral stability, the trailing corners of both wings were twisted simultaneously by cables. In this front view, the right rear corners are twisted downward and the left rear corners upward. Wing warping (later, hinged ailerons) also made steering practicable, permitting a roll about the fore-and-aft axis as the rudder was turned.[2]

is not to record "laws of nature" but to state relations among measurable properties—length, weight, temperature, velocity, and the like—to permit a technological system to be analyzed mathematically.

The engineering sciences also differ from pure science in that they have an array of abstract concepts, independent of science, that serve as a framework within which technical problems can be analyzed. The idea of a thermodynamic cycle is one such concept. In an internal-combustion engine, for example, the working fluid—a mixture of air and fuel—undergoes a series of processes (intake, compression, ignition, exhaust), which can be approximated in a simplified "ideal" cycle for purposes of mathematical calculation. Using the ideal cycle, one can calculate, for example, the "thermal efficiency"—again a technical, not a scientific concept—of a cycle.[16]

An "ideal" efficiency can be useful, but informed judgment must decide to what extent calculations involving idealized processes can be depended upon in making design decisions. Structural analyses (indeed, any engineering calculations) must be employed with caution and judgment, because mathematical models are always less complex than actual structures, processes, or machines.[17]

Design as Invention

To design is to invent. Anyone who pays attention to the way the built world has been put together has probably wondered why a particular design was adopted rather than a reasonably obvious alternative. As one thinks about the essential nature of an alternative design, one mentally formulates an invention, usually employing familiar elements in a new combination to accomplish a particular purpose.

My first mental formulation of an invention was made in 1937 upon seeing a Douglas DC-3 land. As the airplane touched down, the tires screeched and the plane bucked as the sudden friction of the runway brought the wheels up to landing speed. Until that moment I did not know that a problem existed, but on the spot I envisioned a landing-wheel spinner that would prevent what to me was the obvious danger of the tires' blowing out. I assumed it would be straightforward to use an electric or hydraulic motor to spin the wheels. After writing to the Douglas Aircraft Company to announce my invention, I received a friendly letter from a vice-president—no doubt the one in charge of crank letters—informing me that, yes, there were several ways to spin the wheels to get them up to landing speed, but all required weight that would be better spent in making the tire treads thicker. At that point, I let my idea die. A few years later, in 1945, a B. F. Goodrich advertisement displayed a picture of an airplane tire with flaps on the sidewalls; the flaps were opened by the air rushing past and the landing wheels were brought to a speed approaching landing speed. After explaining the problem, the advertisement asserted: "Any engineer could think of an answer—make the wheels spin as the plane comes in. People even had a name for it—'prerotation.' Yet no one found a way to make prerotation practical until B. F. Goodrich designed the tire above."[18] Apparently that idea died nearly as quickly as mine. There has been more talk of prerotation, but today when an airplane touches down the tires still screech, the plane still bucks, and blowouts are still a danger, despite heavy-duty "42-ply-equivalent" tires.

The above tale, trivial in itself, suggests that every technical problem has alternative solutions—often several. One that was not obvious to me was the Douglas solution: to reject solutions that required more machinery and to use the weight saved to make a simpler system more robust.

My idea of a wheel spinner is an example of what might be called *low-level inventing*. It is mechanical design, the sort of thing that technologists do readily and habitually. Design and invention lie along a continuum that ranges from the obvious to the inspired, from design routines that involve a minimum of intellectual engagement to original, fundamental inventions that change forever our way of tackling certain problems. The philosopher Carl Mitcham gives design and invention their proper places in the scheme of things by observing that "invention causes things to come into existence from ideas, makes world conform to thought; whereas science, by deriving ideas from observation, makes thought conform to existence."[19]

One fundamental invention—an inspired *tour de force* of engineering design—was Thomas Newcomen's steam engine (figure 1.6), whose final configuration was worked out over a period of several years preceding its first successful operation in 1712.[20] It is a valuable example that underlines the fallacy of accepting the notion that form follows function, an idea that popular writers have widely and uncritically accepted as a principle of design.

Newcomen designed his steam engine to power a water pump located at the bottom of a coal mine. Except for the specification that water was to be removed from the bottom of a mine, there was no "function" for form to follow. There was little to tell Newcomen what his engine should look like, what the elements should be, or how they should be arranged.

Newcomen's water pump was of unprecedented size, but the configuration of a reciprocating pump was well known and posed no insurmountable problems of adaptation. On the other hand, his was the first steam engine: there was no prototypical machine to adapt. Every feature of his design had to be chosen *de novo*. His extraordinary achievement is not diminished by the fact that a series of essential principles embodied in experimental apparatus had been illustrated in books and journals over the 50 years preceding his work.

The Nature of Engineering Design

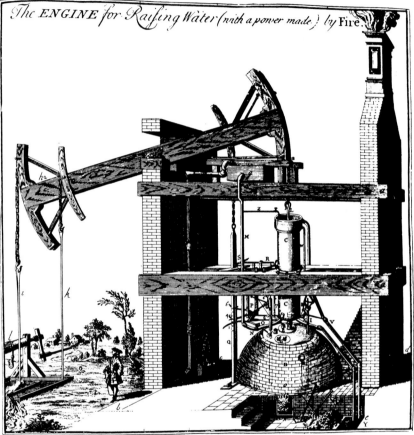

Figure 1.6
Newcomen steam engine and water pump, 1717. This is the earliest known drawing of a Newcomen engine. These enormous machines (see man in left foreground) were first used to remove water from coal mines in England. Within 20 years after the first such engine was built, in 1712, more than 100 were at work throughout Europe.[3]

Newcomen's engine employed a vertical steam cylinder under one end of a massive overhead working beam, pivoted at its center, and a vertical water pump under its other end. Steam from a hemispherical boiler was piped to the steam cylinder. When the cylinder was full of steam, water was injected and the steam was condensed, thus producing a partial vacuum in the cylinder and permitting atmospheric pressure to push the piston downward. Attached to the beam's end by a chain, the piston pulled one end of the beam down and provided power to operate the water pump. More steam was led into the cylinder, the piston rose, and the operations of the engine continued in sequence. The pistons in both the steam cylinder and the water-pump cylinder moved up and down in straight lines because the supporting chains were wrapped and unwrapped around circular arcs at the ends of the working beam.

The principles of Newcomen's engine can be derived from the demonstrations of Otto von Guericke and Denis Papin, and the sector and chain were borrowed from the *Mémoires* of the Paris Academy of Sciences (see figures 1.7–1.12). But no one was there to tell Newcomen how to choose or arrange the elements. He was faced with a difficult design problem; he is remembered because he possessed the mystical gift of definitive design.

Newcomen's design of a steam engine was so obviously correct that it remained essentially unchanged for over 60 years. In the 1770s James Watt made important alterations in the way steam was used to produce power and deliver it from an engine, yet his steam engine retained the essential form of the Newcomen engine—with a vertical steam power cylinder under one end of a heavy working beam and a water pump or connecting rod suspended from the other end of the beam. Newcomen's brilliant solution of the problem of providing straight-line motion of pistons in vertical cylinders is still employed in the "horsehead" rigs used to pump oil from wells around the world (figure 1.13).

The limits of any design are culture-bound: all successful designs rest solidly on specific precedents. When designers are thinking out their preliminary designs, their visual memories are particularly influential. Mark Clark has found significant visual similarities between components of the docking mechanisms of spacecraft and the landing gear of airplanes. The designers

16

Figure 1.7
Steps toward the Newcomen atmospheric steam engine, I. The atmosphere pushes a
piston into a partial vacuum, producing work (defined as a force acting through a distance). The man at right evacuates the cylinder with an air pump. The piston in the
cylinder moves downward, overcoming the resistance of men holding ropes.[4]

Chapter 1

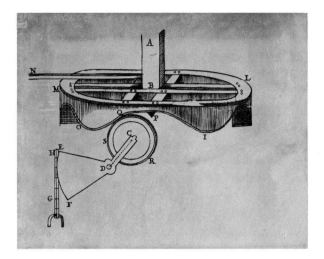

Figure 1.8
Steps toward the Newcomen steam engine, II. The steam piston and the pump rods of
the Newcomen engine were constrained to move vertically by chains wrapped on arcs
at ends of the working beam. This drawing was published in 1695 as the power train
of a "virtually frictionless" water pump. The flat chain HG wraps around the sector
arm DEF and moves the piston rod of the water pump, attached to the yoke near G,
up and down. The scalloped cam, which pivots at F, is turned about its vertical axis
by a horse pulling the arm extending to the left at N.[5]

of the docking mechanisms had been recruited from the pool of aircraft
designers who had had a great deal of experience with landing gear. The
similarities, often subtle, became evident to Clark only as he became familiar
with both airplane landing gear and docking mechanisms. Clark found that
the Russian Soyuz system was radically different from the American Apollo
system. It was arguably easier to operate, and it used electrical instead of
hydraulic components, thus avoiding the hazards of freezing and leakage of
the hydraulic fluid. The Russians and the Americans built full-size mockups
and produced films to persuade each other of the superiority of their respec-

The Nature of Engineering Design

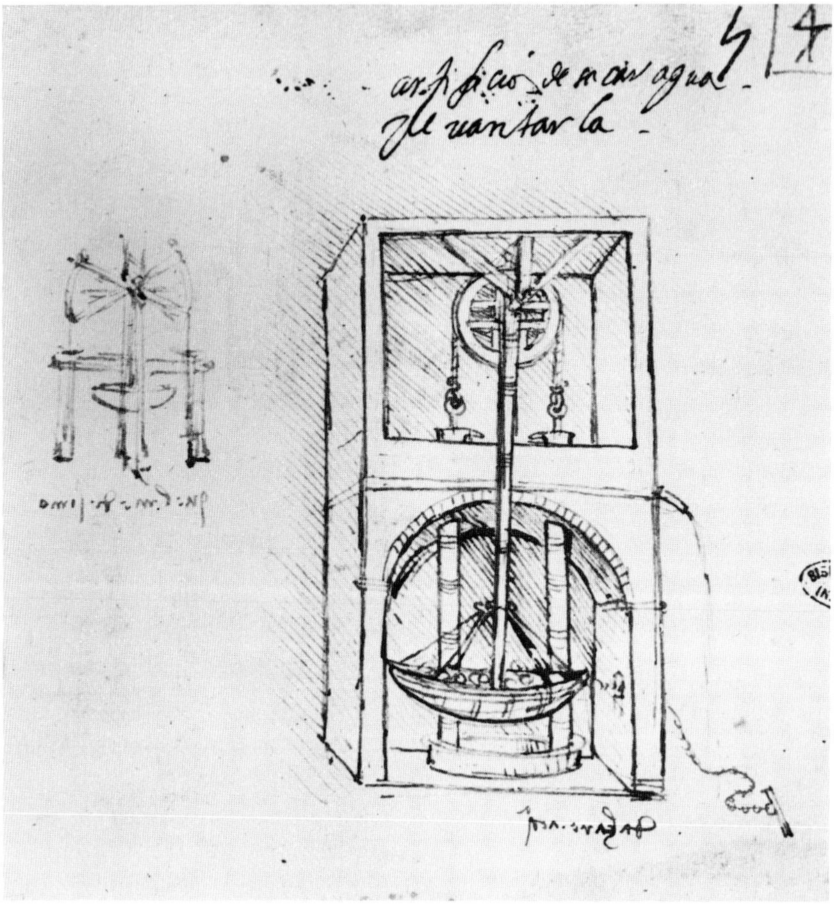

Figure 1.9
Steps toward the Newcomen steam engine, III. Leonardo's two-cylinder water pump
(c. 1500) included a boat-shaped pendulum attached to a wheel overhead. At left,
Leonardo substituted for the wheel a pair of sector arms. Beneath the sketch are the
words "quel medesimo" ("the same").

Figure 1.10
Steps toward the Newcomen steam engine, IV. In Leonardo's gunpowder engine
(c. 1500), gunpowder is placed on pan near the bottom of a cylinder; when ignited, it
heats the air and expels most of it from the cylinder. If the flat cap is quickly clapped
over the mouth of the cylinder, pressure in the cylinder drops below atmospheric, the
piston (attached by a rod to a rectangular weight) rises; if wedges are instantly put in
place, the weight remains at its higher level.[17]

tive approaches. In the Apollo-Soyuz mission of 1975, when Soyuz and
Apollo vehicles were docked in orbit, the only features they had in common
were the external mating elements.[21]

A clearly evident similarity between a designer's visual memory and the
object he designs can be seen in the proposals for a space station submitted to
NASA in 1961 by the Goodyear Tire Company (figures 1.14, 1.15). Clark
notes that Goodyear's "early designs for an inflatable space station look
exactly like an enormous bias-belted automobile tire."[22]

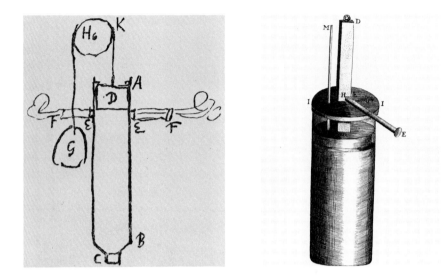

Figure 1.11.
Steps toward the Newcomen steam engine, V. Christiaan Huygens' 1673 gunpowder engine is an improvement on Leonardo's but employs the same principle.[8]

Figure 1.12
Steps toward the Newcomen steam engine, VI. In Denis Papin's steam cylinder (1690), water was evaporated in a cylinder and condensed by dashing water over it. This steam cylinder, whose piston was pushed down by the atmosphere when steam condensed, was described in a journal published in 1690.[9]

A 1924 proposal by Joseph B. Strauss for the Golden Gate Bridge carried the unmistakable stamp of the designer's career as a leading builder of bascule drawbridges—bridges whose counterweights lifted the draw spans by rolling action (figures 1.16, 1.17). Strauss campaigned unsuccessfully for financial and political support for eight years. When he shifted to a conventional suspension span, support was forthcoming, and Strauss was made chief engineer.[23]

Figure 1.13
Horsehead oil-pumping rig in Oklahoma, 1973. Driven by the internal-combustion engine at extreme left, the T-shaped crank arm lowers and raises the left end of the
working beam. The Newcomen-inspired sector at right raises and lowers a cable operating a pump deep in the well below the rig.[10]

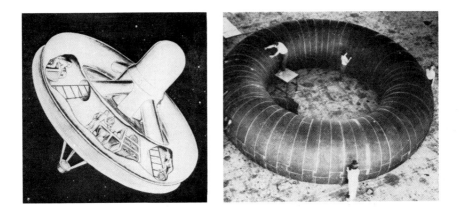

Figure 1.14
This Goodyear Tire Co. proposal for an inflatable space station epitomizes the culture-bound limitations of design. It was to be "folded within a rocket nose cap" and inflated in space.

Figure 1.15
The full-size mock-up of the Goodyear space station demonstrates how much the concept owed, as Mark Clark suggests, to a bias-belted automobile tire.[11]

The historian George Basalla argues convincingly that the designers of new technology cannot ignore old technology, even though they may try to do so. Because inventors and designers nearly always devise new combinations of familiar elements to accomplish novel results, links to known technology are inevitably present.[24] The inevitability of the old in the new is no check to originality, however. The possible combinations of known elements exceed comprehension, and the combination of elements is subject to endless variation.

Art and Engineering
Most engineers today are happy to be called scientists but resist being called artists. Art, as it is understood in engineering schools, is effete, marginal,

and perhaps useless. It is a "soft" subject, lacking the rigor of the hard sciences and the supposed objectivity of engineering.

Yet engineers' drawings, whether made with pencils and pens on a drawing board or with an electronic cursor on a computer screen, share important characteristics with the drawings and paintings of artists. Both the engineer and the artist start with a blank page. Each will transfer to it the vision in his mind's eye. The choices made by artists as they construct their pictures may appear to be quite arbitrary, but those choices are guided by the goal of transmitting their visions, complete with insights and meaning, to other minds. And an artist, other than a particularly anarchic one, generally follows rules implicit in a particular period and a specific style or school.

The engineer's goal of producing a drawing of a device—a machine or structure or system—may seem to rule out most if not all arbitrary choices. Yet engineering design is surprisingly open-ended. A goal may be reached by many, many different paths, some of which are better than others but none of which is in all respects the one best way.

Design engineers have recourse to analytical calculations to assist them in making decisions, but the number of decisions that are based on intuition, a sense of fitness, and personal preference made in the course of working out a particular design is probably equal to the number of artists' decisions that engineers call arbitrary, whimsical, and undisciplined.

On the basis of his experience in training industrial designers, David Pye observes that one who "is capable of invention as an artist is commonly capable also of useful invention."[25] Leonardo da Vinci may be an exceptionally strong example of the combination of artistic and practical talents, which Pye suspects "are really different expressions of one potentiality," but he is a member of a very large group of people, as nineteenth-century American examples illustrate. Benjamin Henry Latrobe (1764–1820), a prominent consulting engineer and architect, was an accomplished watercolorist. Robert Fulton (of steamboat fame) and Samuel Morse (inventor of the electrical telegraph) were both professional artists before they turned to careers in technology.[26]

The reinforced concrete bridges designed by Robert Maillart, a Swiss engineer who graduated in 1894 from the Swiss Federal Technical University

Figure 1.16
The Golden Gate bridge that might have been. Proposed in 1924 by Joseph B. Strauss, a leading builder of bascule (rolling) drawbridges, this design suggested two huge bascule bridges connected by a suspension span. The towers were to be 4000 feet from center to center.[f12]

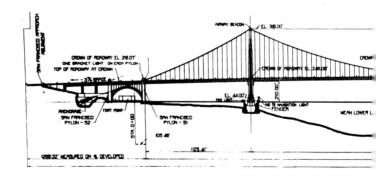

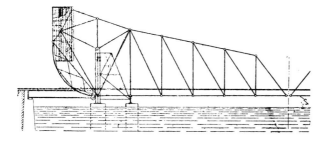

Figure 1.17
Patent drawing of bascule drawbridge patented by Joseph B. Strauss (US Patent 1,150,643, issued August 17, 1915).[13]

Figure 1.18
The Golden Gate Bridge as completed in 1937. The center span is 4200 feet.[14]

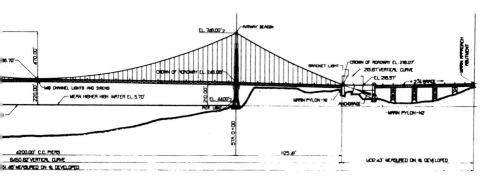

The Nature of Engineering Design

in Zurich, epitomize the technical and aesthetic aspects of what David Billington, a professor of civil engineering at Princeton, calls "structural art." According to Billington, structural art began in the Industrial Revolution, is "parallel to and fully independent from architecture," and is being practiced in the twentieth century by "engineering artists."[27]

Maillart's bridges, strikingly original in form, established a tradition of choosing structural shapes in which forces could be so readily visualized that mathematical analysis was reduced to a minimum. Load testing of finished bridges and close and continuing observation of their performance provided both quantitative and qualitative knowledge that affected future choices. Maillart chose the form of a bridge in such a way that the forces followed the form. In effect, he arranged the load-bearing members of his bridges while building in his mind a force diagram that would permit a graphical analysis of stresses.

In one of the first bridges he built, in 1901, a series of cracks developed in the arch walls near the abutments. Maillart watched the bridge carefully and eventually concluded that the arch wall in the region of the cracks was unnecessary for support. In a bridge he designed in 1904, he omitted the section that had cracked and produced, in Billington's words, "a new form with unprecedented visual power, increased material efficiency, and decreased cost for construction and maintenance—in short, a better bridge"[28] (figures 1.19, 1.20).

An Engineer's Style

Maillart's bridges exhibit the personal "style" of their designer, embodied in his early reinforced concrete hollow box arches and refined through subsequent insights that permitted more economical and elegant forms.

The major inventions of Thomas Edison also exhibit a style, but one that is eclectic rather than exclusively personal. The historian Reese Jenkins, who perceived a visual thread connecting Edison's various inventions, described that thread as composed of "elements of style."[29] Edison used again and again an array of mechanical combinations that he adapted to such diverse machines as the phonograph, the printing telegraph, the electro-

Figure 1.19
Robert Maillart's 1901 bridge over the Inn River at Zuos, Switzerland (near St. Moritz). The 125-foot bridge was the first of Maillart's influential and totally original reinforced concrete bridges, with a deck, a flat arch, and sidewalls forming a box beam of varying depth.[f15]

Figure 1.20
Because vertical cracks, not structurally dangerous, appeared after two years in the sidewalls near the abutments of the Zuos Bridge, Maillart eliminated segments of the sidewalls in his 1905 Rhine River Bridge at Tavanasa. The result, in Billington's words, was "a new form with unprecedented visual power."[f16]

mechanical telautograph (which reproduced at a distance a written input), and the kinetoscope (predecessor of motion pictures).

One combination of elements of style that appeared in several of Edison's designs was a rotating drum or cylinder, generally mounted horizontally, whose surface was sensitive to a stylus. Although a flat-turntable machine appears among Edison's sketches of ideas for his phonograph, he preferred a rotating cylinder and built his system around it (figures 1.21, 1.22). He used a narrow treated ribbon or paper tape for the recording surface in his printing telegraph and stock ticker, and he considered using the same principle in one version of his phonograph. Jenkins concludes that "a creative technologist possesses a mental set of stock solutions from which he draws in addressing problems."[30]

A "stylistic analysis," advocated by Jenkins, can often identify the "elements of style" and "stock solutions" that give a distinctive "family" resemblance to diverse technological artifacts. The machines, structures, and devices of the Victorian period, for example, can often be identified and dated by one who has developed intimate familiarity with a wide range of Victorian examples. The diversity that is found in a particular style again points up the wide range of acceptable solutions that a given design problem can elicit.

The Process of Design

In principle, the process of design and construction is the same no matter how small or large, how simple or complex. A simple, routine design, such as that for a concrete pedestal on which to mount a heavy machine in a job shop, may merit a sheet or two of drawings that specify the shape and the dimensions of the concrete, the location and size of the holding-down bolts, and perhaps the location of an electrical conduit. The construction of the pedestal may be carried out by the shop's maintenance crew or be farmed out to a small local contractor. In either case, if the drawing is accurate and has been followed by the builders, the holding-down bolts can be expected to fit the machine and the electrician will have a clear run for the wires. A complex project, on the other hand, may require thousands of drawings.

The Tarbela dam and power plant, on the upper reaches of the Indus River in Pakistan, was designed by a New York engineering and construction firm. Initially, 500 contract drawings and three volumes of specifications were required to ensure an understanding by all parties of what was to be built and exactly where. In addition, some 3600 construction drawings were made in the home office, and twice as many in the field.[31]

Although the planning of a large engineering project such as the Tarbela dam may employ hundreds of engineers, many of whom will have detailed knowledge and skills not possessed by the engineer in charge, to be successful the entire design must be always held in the chief engineer's mind. That

Figure 1.21
Edison's inventive "style" favored a phonograph with a rotating drum, mounted horizontally, whose surface of tinfoil was sensitive to a stylus. While the cylindrical tinfoil
"record" was rotated by turning the crank, the diaphragm at B recorded sounds. The
same diaphragm also acted as a pickup, transmitting sound to the speaker cone.[17]

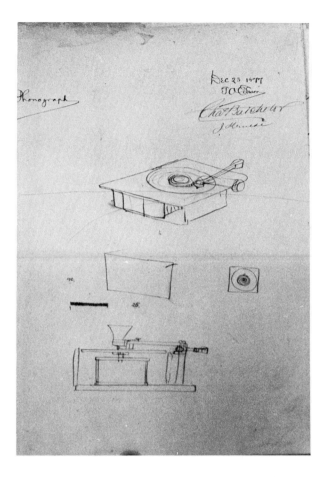

Figure 1.22
Although he preferred the arrangement shown in figure 1.21 for his commercial pho-
nographs, Edison had also considered a disc-shaped "plate." He proposed to reproduce
"music both orchestral, instrumental, and vocal, the idea being to use a plate machine
with perfect registration & stamp the music out in a press from a die or punch. . . ."

image should include sufficient detail to ensure that the various components perform harmoniously and that crucial features and details will not drop through the cracks between adjacent specialist designers' desks. All kinds of flow charts and other remembering devices are used, because formal paper systems are often more dependable than human memories in reminding all hands of schedules and deadlines; however, experienced engineers understand that their minds are more likely than bureaucratic systems to notice faint indications that a major unexpected problem will develop. Because every project accumulates a great deal of tacit information and many tacit understandings, the chief engineer must have uncommon ability to ensure that the finished design describes a product that will operate as expected without excessive adjustments.

The need for this kind of deep knowledge[32] and understanding of a project was demonstrated in an extreme way in 1970 when the command module of Apollo 13 was damaged some 210,000 miles from Earth by the bursting of two oxygen tanks. The chief designers of the command module and of the landing module were flown to the Mission Control Center in Houston to provide immediate diagnoses, judgments, and recommendations. The damage precipitated an emergency that grew into a harrowing three-day exercise in troubleshooting and in adapting available means to compensate for unimagined shortcomings in the spacecraft's design.[33]

Those who observe the process of engineering design find that it is not a totally formal affair, and that drawings and specifications come into existence as a result of a social process. The various members of a design group can be expected to have divergent views of the most desirable ways to accomplish the design they are working on. As Louis Bucciarelli, an engineering professor who has observed engineering designers at work, points out, informal negotiations, discussions, laughter, gossip, and banter among members of a design group often have a leavening effect on the outcome.[34]

James Althouse, a designer of service equipment for oil wells, has recounted the way one of his designs (figure 1.23) was influenced by an informal discussion at lunch with several other engineers. They agreed that commercially available liquid-level controllers were unreliable when used to maintain a constant level in a highly agitated truck-mounted tank that dis-

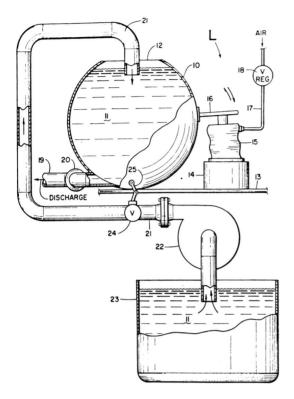

Figure 1.23

Was this patent steered by lunchtime banter? The patent, by James W. Althouse III, is for "an apparatus and method for controlling the level of a liquid in a tank." The upper tank is filled by a centrifugal pump (22) taking its supply from a reservoir (23). The upper tank is supported by a stable platform (13)—actually, the bed of a motor truck—and an "air spring" (15); when the water level drops, the air spring expands and rolls the tank a few degrees counterclockwise. The tank's movement opens valve V (24) and the centrifugal pump refills it, whereupon the air spring is compressed, the tank rolls slightly to the right, and the valve is closed when the tank is full. The apparatus was designed to keep a constant level in the upper tank, which supplies huge quantities of a water-clay ("mud") mixture for use in oil-well maintenance.

The Nature of Engineering Design

charged a high volume of liquid into a well while being refilled from a stationary reservoir. As Althouse recalled the event:

> Someone said it was like trying to keep a bucket with a hole in the bottom full without splashing with an open-ended garden hose turned on full blast. Someone else said that if you kink the hose you can regulate the flow pretty well by feel. Someone else suggested that we train giant gorillas to kink a 4-inch high pressure hose and watch the tank level. . . .
>
> Thinking about the problem the same afternoon, I remembered that I had once used a filled bucket to hold the kink in a hose to keep it from spraying until I could go back and shut off the valve. The bucket had to be full or it wouldn't hold the kink. Out of nowhere it occurred to me that if the mixing tank were allowed to pivot on its outlet piping it could regulate its own fill-up valve with a direct linkage—kind of like sitting on the fill-up hose. All we needed was to counter-spring the weight of the tank so we could use a conventional valve. Sketched that afternoon, I got a relatively uncontested patent on the device. I am not claiming . . . that nobody else would think of it if they were familiar with the strange circumstances of mobile processing equipment design. What is interesting to me is whether or not I would have thought of it if the usual lunchtime banter had taken a different tack.[35]

A brief review of engineering design as it was done for centuries, until after 1970, may suggest to different readers different traditional elements that have been lost to those who inhabit workstations. The accompanying photograph of the Baltimore & Ohio Railroad drafting room (figure 1.24) shows where and how locomotives, rolling stock, and railroad structures were designed. The designer in the right foreground appears to be making a sketch that will be converted by a draftsman into a working drawing. He has for immediate reference papers, sheafs and rolls of blueprints, and printed materials. At the far end of the room, original full-size drawings are stored flat in drawers for immediate retrieval. Data books and sketch books are in the foreground. A consultation between two colleagues is taking place at the center of the picture. The results of the designers' and the draftsmen's work are displayed in framed photographs on the walls, and one can walk out into the yards to see the real thing. The designers are thus intimately in

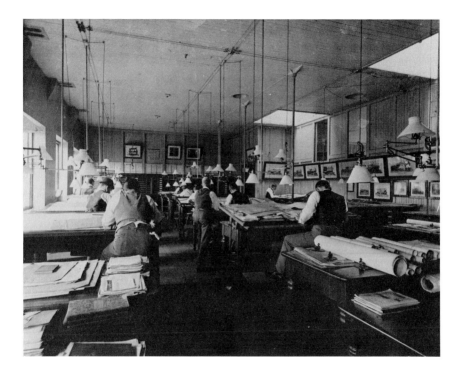

Figure 1.24
Drafting room of Baltimore & Ohio Railroad, Baltimore, 1899.[f18]

touch with the world they have designed, and they are engaged intellectually and physically at a detailed level in planning the future of their railroad.

In the late 1950s, when instruction in engineering drawing began to disappear from engineering schools, an occasional suggestion was made that finished drawings were not really necessary to convey a designer's instructions to the shops. Perhaps sketches would suffice. Although such suggestions were not taken seriously by many engineers, profound changes in converting designers' ideas to shop instructions were in fact under way.

In 1961, a radical new system of designing and drafting called "Panoramic Design Technique (PDT)" was developed by TAB Engineers, Incorporated, a small Chicago firm. As though straining to eliminate everything that made a drafting room a vital place to work, the PDT system banished drafting tables, put drawings on blackboards, and made a record of drawings by photographing the blackboards. Figure 1.25 conveys the intellectual impoverishment of the new system. Continuous discussion in each group (in order to appear busy) eliminated contemplation, the time to think through a promising concept, to refer to other drawings, notebooks, and printed books. The absence of reference materials and the apparent disdain for any reminder of the company's products ensured that designs would be based on first thoughts and a minimum of time-consuming reflection and rethinking of doubtful concepts. Its promoters claimed that the new system would save one-third to one-half of "design, drafting, and engineering costs" while giv-

Figure 1.25
The PDT (Panoramic Design Technique) program was developed in 1961 by TAB Engineering, Inc. The new method was expected "to cut from 33 to 50 percent from drafting, design, and engineering costs."[f19]

ing management "better control" of the entire process.[36] The badly conceived PDT system was quickly forgotten, but by 1965 computer-based systems of automatic drafting (using punched cards) were being promoted by computer manufacturers and computer-aided design systems were being developed in a few engineering schools.[37]

Engineering design is always a *contingent* process, subject to unforeseen complications and influences as the design develops. The precise outcome of the process cannot be deduced from its initial goal. Design is not, as some textbooks would have us believe, a formal, sequential process that can be summarized in a block diagram (figure 1.26). Starting with a block called NEED, such a block diagram—which may comprise from a dozen to more than a hundred blocks—purports to guide (or at least to follow) the designer through the process of inventing and analyzing a new thing. Block diagrams imply division of design into discrete segments, each of which can be "processed" before one turns to the next. Although many designers believe that design should work this way, even if it doesn't,[38] it is clear that any orderly pattern is quite unlike the usual chaotic growth of a design. The vision at the heart of a design is often in a designer's mind long before a need has been articulated. Second thoughts are admitted in the block diagrams along "feedback" paths, but the reader should understand that the steps in the design process may all be going on at once.

Despite its complexity, subtlety, and refusal to fit into neat diagrams, however, engineering design follows a predictable path whose nature will not be changed by computer-assisted design (CAD) or by a wished-for science of design. Computerized illusions of certainty do not reduce the quantity or the quality of human judgment required in successful design. To accomplish a design of any considerable complexity—a passenger elevator or a railroad locomotive or a large heat exchanger in an acid plant—requires a continuous stream of calculations, judgments, and compromises that should only be made by engineers experienced in the kind of system being designed. The "big" decisions obviously should be based on intimate, firsthand, internalized knowledge of elevators, locomotives, or heat exchangers. Every designer ought to have an intuitive sense of the practical limits of the

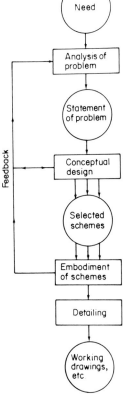

Figure 1.26
A block diagram (which many people call a flow diagram) embodies the engineering conviction that any problem can be solved if only it can be broken down into enough parts, or steps. This diagram illustrates M. J. French's idealization of the design process.[f20]

performance of moving machinery and a broad sense of the adequacy of materials and of fabrication processes.

Design layouts and calculations require dozens of small decisions and hundreds of tiny ones, because the topography of a new thing is astonishingly unconstrained and because numerical calculations always embody human judgment. In setting up calculations, for example, one must make assumptions about the initial conditions and the exponents of the process curves. Though none of those minor choices may appear to be decisive in the completed calculations, they do form an intricate but hidden foundation of the design.[39]

If designers use commercial computer programs, which are becoming available for more and more classes of problems, they turn over all the small and tiny decisions to the programmer, who is more likely to be an "engineering scientist" than an experienced designer. To unearth all the points of judgment and decision in an extensive computer program is difficult if not impossible. Yet those small decisions can be fatal to the success of the design. For example, the space-truss design for the roof of the Hartford Coliseum included more than 200 slender cruciform bars, each 30 feet long, whose strength in compression had been calculated. The roof failed under a moderate snow load because some of the long compression members buckled and brought the rest of the truss down in domino fashion. The programmer apparently had not expected those long members to be subjected to anything but pure compression. The possibility that a partial roof collapse might cause one or more members to buckle and thus nullify most of the assumptions made by the programmer either was not considered or was judged to be too remote to warrant inserting the several hundred stiffening braces necessary to arrest the domino action of an unbraced truss. Now this is a small decision in the scheme of things, although it will no doubt be a big consideration in the design of future space trusses. If somebody involved in the Hartford design had seen or had been able to visualize some of the buckling accidents that have occurred since the 1907 collapse of a railroad bridge under construction at Quebec City, stiffening stays or braces might have been added. In any case, assumptions and matters of judgment will always be present in engineering design, whatever the format of the design. Because not all assumptions can be made explicit—there is too much tacit knowledge and too many inarticulate (and inarticulable) judgments to make that possible—it is important to put the assumptions, judgments, and decisions (big, small, and tiny) in the hands of designers who have studied reality as well as the engineering sciences. As David Billington counsels, "All engineers of the late twentieth century need to know the computer well; all designers need to keep from relying on it for their basic structural experience."[40] That experience must be obtained in the field, where the structures are. Competent structural designers are familiar at first hand with the construction as well as performance of those structures. Designers of mechanical and hydraulic systems must have field experience appropriate to their needs, as must all successful engineering designers.

2 The Mind's Eye

One time, we were discussing something—we must have been eleven or twelve at the time—and I said, "But thinking is nothing but talking to yourself."

"Oh, yeah?" Bennie said. "Do you know the crazy shape of the crankshaft in a car?"

"Yeah, what of it?"

"Good. Now tell me: how did you describe it when you were talking to yourself?"

So I learned from Bennie that thoughts can be visual as well as verbal.

Richard P. Feynman, 1988[1]

. . . and when needful [the hinged railings] are stood upright as is made clear and obvious in the illustration because I cannot so well set it forth in words as I see it in my mind's eye. But the picture will show it.

Guido da Vigevano, 1335[2]

Everyone, with the possible exception of those who do not dream, is familiar with images in the mind. Many of us resort without reflection to our nonverbal abilities to think in images of real things and of things that appear only in our imaginations. The ability to visualize is so widespread that many of us are surprised and a bit skeptical when we are told by anyone that he or she never uses imagery in thinking.

Visual thinking is necessary in engineering. A major portion of engineering information is recorded and transmitted in a visual language that is in effect the *lingua franca* of engineers in the modern world. It is the language that permits "readers" of technologically explicit and detailed drawings to visualize the forms, the proportions, and the interrelationships of the elements that make up the object depicted. It is the language in which designers explain to makers exactly what they want them to construct.

The mind's eye, the locus of our images of remembered reality and imagined contrivance, is an organ of incredible capacity and subtlety. Collecting and interpreting much more than the information that enters through the optical eyes, the mind's eye is the organ in which a lifetime of sensory information—visual, tactile, muscular, visceral, aural, olfactory, and gustatory—is stored, interconnected, and interrelated.[3] We get to know things through a series of sensual interactions: bumping, smashing, touching, smelling, dropping, lifting, and so on. The arbiter of all these experiences is the mind's eye. Through it, we make sense of the physical world we inhabit. When we see a piece of cloth lying on a chair we can say without touching it that it is probably soft (or harsh), pliable (or stiff), and a good (or poor) insulator against cold. A tailor who has used that kind of cloth can add a whole new order of facts and insights regarding its attributes.

Some of the operations that the mind's eye can carry out are impressive, and some are breathtaking. The mundane act of driving a car past another vehicle on a busy highway involves instant, unreflective judgments of the positions, relative speeds, and bearings of a large number of objects, moving and fixed. The task of finding one's location on a small, flat map requires the mind's eye to modify the shapes sensed in the three dimensions of reality in order to reconcile what the optical eye sees and what the map asserts. Visual thinking can be successful to the extent that the thinker possesses an adequate array of sensual experience, converted by the mind's eye to usable visual information.

The Right Brain and the Left Brain

Physically, the mind's eye is supposed by many authors to be on the right side of the brain. That some people are left-brained and others right-brained is a popular dogma. According to the journalist Maya Pines, the right side of the brain is the seat of "artistic [and] musical ability [and] spatial perception," while the left side of the brain is the locus of "language and analytical ability."[4] *Readers Digest* jumped on the bandwagon with a superficial article entitled "Are You Thinking Right?"[5]

Roger W. Sperry, who in 1981 was awarded a Nobel Prize for his studies of brain trauma involving the separation of the cerebral hemispheres, recalled that within his lifetime the "so-called subordinate or minor hemisphere" was supposed to be "illiterate and mentally retarded and thought by some authorities not even to be conscious."[6] Pines, reflecting those beliefs, declares in her book *The Brain Changers* that the right half-brain is not stupid but "merely speechless and illiterate. It actually perceives, feels, and thinks in ways of its own, which in some cases may be superior. The only problem is to communicate with it nonverbally, as if it were an exceedingly intelligent animal."[7] However, in Sperry's experience, the right hemisphere, when disconnected from the left, apparently draws upon "residual abilities" and is capable of cognitive capacities once ascribed exclusively to the left hemisphere. He suspects that those "residual" faculties had not been recognized in earlier cases of damage to the left hemisphere because the two halves had not been completely disconnected, and he sounds a note of caution about the growing use of semi-popular extrapolations and speculations concerning "left-brain" and "right-brain" functions: "The left-right dichotomy is an idea . . . with which it is very easy to run wild."[8]

Around 1900, when their discipline first aspired to the status of a science, American psychologists considered mental operations to be keys to behavior. They studied imagery, which they believed to be visual, auditory, tactile, or purely verbal. Some analysts were particularly interested in pathological images, such as hallucinations. Although important insights were gained (for example, that a person may solve a problem without being aware of the process followed in the mind), mainstream psychology soon discarded "mentalistic" studies and adopted the "hard" science of behaviorism, in which introspective reports of what went on in a subject's mind were banished in favor of objective scientific testing of a subject's behavior under conditions set by the experimenter. Because images could not be measured, they were declared irrelevant.

In the 1950s, however, the study of imagery became respectable again when practical problems in engineering psychology demanded solutions. These problems included how to deal with the vivid imagery—sometimes mistaken for reality—that troubled radar operators, long-distance truck drivers, jet pilots, and operators of snowcats in blizzards. By the 1980s, imagery had become "one of the hottest topics in cognitive science."[9]

The Status of Visual Thinking

In 1880, when Francis Galton, founder of the "science" of eugenics, studied the intellectual characteristics of prominent British scientists, he was astonished to find that the thought processes of most of them were quite different from his own. Galton thought in visual images, he said, while the majority of scientists reported that they thought in words, with seldom any suggestion of an image. In deference to his subjects, most of whom were of "very high repute," he conceded that the "visualizing faculty" was probably inferior to "the higher intellectual operations [of words]." Although those scientists might at one time have had the visualizing faculty, Galton said, it had probably been lost by disuse.[10]

Galton was a member of the aristocracy, which took for granted its superior intellectual abilities, but he hesitated to challenge the eminent scientists who "think hard." He may have been a bit humbled when he learned that the people he met on the unexalted level of "general society" were, like him, visual thinkers. He supposed that "mechanicians, engineers, and architects possess the faculty of seeing mental images with remarkable clearness and precision. . . . inventing their machines as they walk, and see[ing] them in height, breadth, and depth as real objects, and . . . in action." Galton's conclusion that "an over-ready perception of sharp mental pictures is antagonistic to the acquirement of habits of highly generalized and abstract thought" maintained a proper distance between thinkers and doers.[11]

Whatever the hard-thinking scientists of high repute may have told Galton, the tendency of the leading British physical scientists of the nineteenth century—most notably Michael Faraday, Lord Kelvin, and James Clerk Maxwell—to think in terms of models and mechanical analogies is well established.[12]

Pierre Duhem, a French physicist and historian of science, perceived a sharp difference of style between French and British scientists of the nineteenth century. The French were at home with abstract concepts, he declared, but the British were quite unable to cut through a welter of concrete facts and state general principles. British minds required mechanical models to aid their reasoning. Lord Kelvin, Duhem opined, thought the test of understanding of a particular subject was "Can we make a mechanical model of it?" Oliver Lodge's book on electromagnetic theory had so many strings, pulleys, gears, and pipes in its explanatory models that Duhem was driven to observe: "We thought we were entering the tranquil and neatly ordered abode of reason, but we find ourselves in a factory."[13]

Early in the twentieth century, the prominent American philosopher William James remarked that a favorite topic of discussion among his colleagues was "whether thought is possible without language." James said that it was perfectly possible, and that different thinkers think in different ways: "With one, visual images predominate, with others, tactile."[14] Historians of science in the late twentieth century have documented the persistent use of imagery by Ludwig Boltzmann, Albert Einstein, Niels Bohr, Werner Heisenberg, and others as they developed modern physics. Albert Einstein said that he rarely thought in words at all; his visual and "muscular" images had to be translated "laboriously" into conventional verbal and mathematical terms.[15]

In the late 1940s Richard Feynman, a brilliant theoretical physicist, enhanced the power of quantum mechanics by inventing "Feynman diagrams," a visual alternative to a formidable array of equations. Feynman thought that Einstein, in his old age, failed to develop his "unified theory" because he "stopped thinking in concrete physical images and became a manipulator of equations."[16]

In the 1980s the chemist-philosopher Robert S. Root-Bernstein documented the "extracurricular" abilities of more than a hundred prominent scientists of the eighteenth, nineteenth, and twentieth centuries. Most were part-time visual artists, musicians, and poets; a few were composers, writers of fiction, and photographers. Root-Bernstein argues that those visual and artistic proclivities have a distinct bearing on the originality of the scientists. His extensive studies have led him to conclude with certainty that "most eminent scientists agree that nonverbal forms of thought are much more important to their thought than verbal ones."[17]

Yet Galton's unsubstantiated and even refuted nineteenth-century hier-

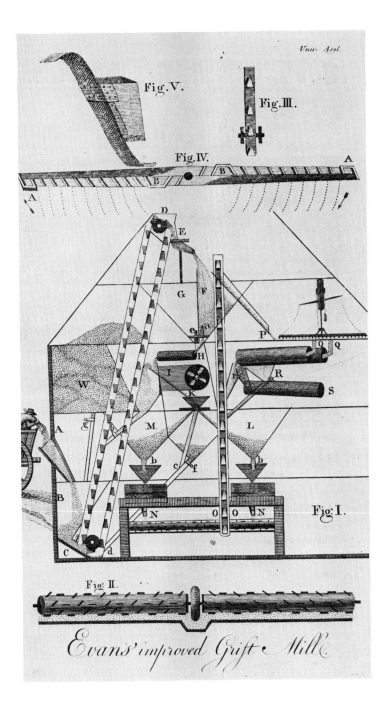

Fig. V.

Fig. III.

Fig. IV.

A

Fig: I.

Fig: II.

Evans' improved Grist Mill.

Chapter 2

archy, with pure verbal "hard thinking" at the top, persists in many academic minds. Telltale phrases pepper the modern writings on cognitive theory: "the visual rather than the purely intellectual aspects of the problem," for example, and "some researchers are eager to give the less intellectual aspects of human personality equal weight with the verbal ones."[18] Some of these expressions are straightforward statements of self-assured patricians; others are windows into scholars' unconscious assumptions of a natural superiority of words over visual images.

The Practice of Visual Thinking

Despite the low academic status of visual thought, it is an intrinsic and inseparable part of engineering. Over the years, a number of creative engineers have revealed their mode of thought as they explained how they solved technical problems.

In the 1780s Oliver Evans, a Delaware farm boy, invented the automatic flour mill, in which bucket elevators and screw conveyors were coordinated to eliminate the need for manual lifting or carrying of grain or flour (figure 2.1). Evans claimed to have first put the system together in his head: "The arrangement I so far completed [in my mind] before I began [to build] my mill that I have in my bed viewed the whole operation with much mental anxiety."[19]

Also in the 1780s, James Watt informed his partner, Matthew Boulton, of an idea he had for a "straight-line" mechanism (to guide a piston rod in a straight line), which became his favorite invention (figures 2.2, 2.3). "I have started a new hare," he wrote; "I have got a glimpse of a method of

Figure 2.1
Oliver Evans' automatic flour mill, 1791. Wheat is moved, before and after grinding, from lower to upper floors in bucket elevators and horizontally in screw conveyors. Evans boasted that "from the time the wheat is emitted from the Waggon [at left] untill it is compleatly manufactured into superfine flour the whole is done by machinery without any part thereof being moved by manual labour."[f21]

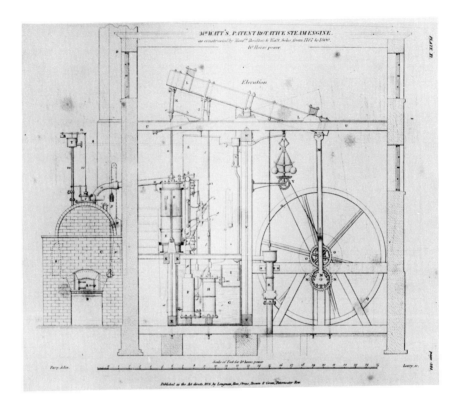

Figure 2.2
James Watt's double-acting rotative steam engine, 1787. The mechanism for the vertical guidance of the upper ends of the steam piston rod and the water pump rod in
"parallel motion" is among the most ingenious of Watt's many original contrivances.[22]

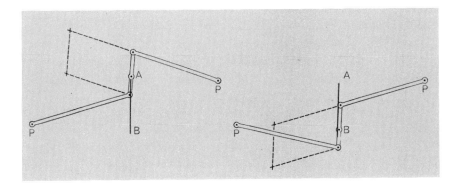

Figure 2.3

Schematic diagrams of Watt's "parallel motion," which kept the piston rod vertical as the beam end moved in an arc. The key elements (solid lines) worked from fixed pivots (P) so that point A, at the center of the vertical element, moved along line A–B. Watt's final version made use of a parallelogram linkage (broken lines) that made the whole apparatus more compact.

causing the piston-rod to move up and down perpendicularly by only fixing it to a piece of iron on the beam, without chains, or perpendicular guides . . . or other pieces of clumsiness."[20] Marc Isambard Brunel, the French refugee engineer who, in England just after 1800, designed semi-automatic machines to make pulley blocks for ships in sequential operations, remarked on the ease with which he expressed his ideas in his drawings; he considered drawing techniques to be the true "alphabet of the engineer."[21]

In describing the origin of his steam pile driver, James Nasmyth, a mid-nineteenth-century English engineer, said that the machine "was in my mind's eye long before I saw it in action." He explained that he could "build up in the mind mechanical structures and set them to work in imagination, and observe beforehand the various details performing the respective functions, as if they were in absolute material form and action."[22]

Walter P. Chrysler, founder of the automobile company, recounted in his autobiography how as an apprentice machinist he had built, without drawings, a model locomotive that existed "within my mind so real, so complete that it seemed to have three dimensions there. . . . My fingers were an intake valve through which my mental reservoir was being filled; of course, my eyes and ears were helping in the process, but what I learned with my fingers and my eyes together I seem never to forget."[23]

Chrysler was the kind of young man Nasmyth would have been proud of. Nasmyth had only scathing words for the young dandies of Victorian England, outfitted in fancy clothes and wearing kid gloves, who called themselves engineers: ". . . the eyes and fingers—*the bare fingers*—are the two principal inlets to trustworthy knowledge in all the materials and operations which the engineer has to deal with. . . . Hence, I have no faith in young engineers who are addicted to wearing gloves. Gloves, especially kid gloves, are perfect non-conductors of technical knowledge."[24]

Michael Pupin, a scientist who had emigrated from Serbia to the United States, wrote of the imaginative originality of Peter Cooper Hewitt, inventor in 1903 of the mercury vapor lamp: "Those who knew him, watching him at work, felt that a part, at least, of Hewitt's thinking apparatus was in his hands."[25]

Elmer Sperry, whose name is tied to the gyrocompass used universally in ships and aircraft and to the gyroscopic stabilizers used in ships, reflected on the circumstances that led to his significant inventions incorporating gyroscopic characteristics and forces:

> First I had somewhat of a library on the gyroscope. Almost without exception these books and pamphlets were terrifying in the profuse use (I have often thought rather in the line of abuse) of higher mathematics. These did not serve me very far, but our family was blessed with three boys and I tried to keep these youngsters supplied with gyroscopic toys of various varieties, some of which I imported. I got more out of these toys than the boys did, inasmuch as they served the very useful purpose of putting me wise as to the magnitudes involved in the gyroscopic reactions that I knew about. These latter were more or less familiar to me, but the former [i.e., the magnitudes] in some respects astonished me. I never would have realized the possibilities had I not been able thus to visualize them while they were actually taking place.[26]

The intensity of Sperry's proclivity for visual thinking was unusual, even among engineers. A veteran employee of the Sperry Gyroscope Company recalled the way Sperry would be "just looking into the air, when all at once he would pick up a pad and hold it at arm's length, then with a pencil in the other hand he would begin to draw. . . . 'It's there! Don't you see it! Just draw a line around what you see.' Whatever he saw he saw 100% perfect, there in the air, but it took a long time and many changes to reproduce in wheels the thing which he saw. He had infinite patience of mind and persistence of will in working out a detail with his engineers."[27]

Whitcomb's Area Rule

One of the most informative accounts of a connected, extended campaign of visual thinking describes how in 1951 a young aeronautical engineer, Richard T. Whitcomb (a 1943 graduate of the Worcester Polytechnic Institute who had gone to work directly after graduation at the Langley Aeronautical Laboratory in Virginia), effectively solved a fundamental aerodynamic problem in the design of supersonic aircraft.[28] Air resistance (drag), which increases with the speed of an airplane, becomes sharply higher as the speed approaches Mach 1, the speed of sound. Increasing the engine power to overcome the drag required much more fuel and thus severely limited the cruising range.

Langley Lab, operated by the National Advisory Committee for Aeronautics (NACA), the predecessor of NASA, employed wind tunnels to provide data needed by designers of aircraft who were faced with problems of lift and drag of wings, vibration and flutter of airframes, cooling of engines, and instrumentation. One of Langley's most valuable early contributions was a series of airfoil designs consisting of scale drawings of wing cross sections (figures 2.4, 2.5). The Langley designs, widely used for all kinds of airplanes, had been developed in wind-tunnel and flight tests.

Whitcomb worked for several years in the "8-foot High-Speed Tunnel," one of several wind tunnels at Langley. In 1945, the power to run the fan supplying air to that particular tunnel had been boosted from 8000 to 16,000 horsepower, thus raising the air velocity from Mach 0.75 to Mach 1; in

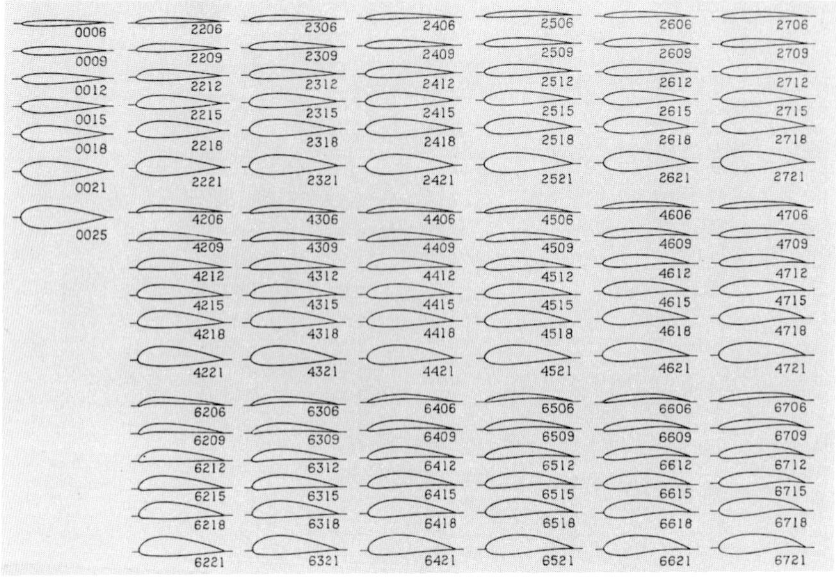

0006	2206	2306	2406	2506	2606	2706
0009	2209	2309	2409	2509	2609	2709
0012	2212	2312	2412	2512	2612	2712
0015	2215	2315	2415	2515	2615	2715
0018	2218	2318	2418	2518	2618	2718
0021	2221	2321	2421	2521	2621	2721
0025	4206	4306	4406	4506	4606	4706
	4209	4309	4409	4509	4609	4709
	4212	4312	4412	4512	4612	4712
	4215	4315	4415	4515	4615	4715
	4218	4318	4418	4518	4618	4718
	4221	4321	4421	4521	4621	4721
	6206	6306	6406	6506	6606	6706
	6209	6309	6409	6509	6609	6709
	6212	6312	6412	6512	6612	6712
	6215	6315	6415	6515	6615	6715
	6218	6318	6418	6518	6618	6718
	6221	6321	6421	6521	6621	6721

Figure 2.4
A catalogue of airfoils for the aircraft industry, published by Langley Laboratory of the National Advisory Committee on Aeronautics in 1933 and based on testing in a variable-density wind tunnel.

1947, changes in the nozzle leading into the throat of the tunnel, where models were suspended for testing, raised the velocity to Mach 1.2, a supersonic speed. In 1948, a slotted throat section, designed to improve the uniformity of flow, was installed, but it took many months of painstaking work by Whitcomb and two colleagues to reach the expected performance. They filed and refiled the slots, guided by close observation of irregularities of air flow and, ultimately, relying on their tactile sensibilities.[29] During those eight years "in the tunnel" Whitcomb was in effect serving his apprenticeship in the mysteries of air flow at high velocities. As apprentice to the wind, he learned a great deal, much of which he could neither verbalize nor

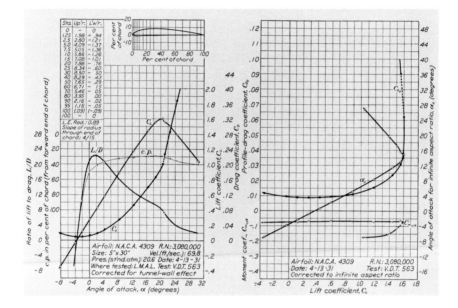

Figure 2.5
The lift and drag characteristics of each of the airfoils shown in figure 2.4 were explained in a series of curves that provided explicit values for use by designers.[23]

describe in any meaningful way to a person whose background did not include similar visual and tactile experience.

In 1949 a "delta" wing configuration was introduced in wind-tunnel experiments to reduce drag on the wings of supersonic aircraft, and a bullet-shaped fuselage was adopted in hopes of decreasing turbulence and thus minimizing drag. When the delta wing and the bullet fuselage were combined in a wind-tunnel model, however, air resistance increased discouragingly as velocity approached Mach 1. Eventually, Whitcomb hit upon a configuration of fuselage and wing that promised to reduce drag considerably. Drag was highest where the wings met the fuselage, which was also (Whitcomb finally realized) where the cross-sectional area of the aircraft was a maximum. One day, in his office, the solution slipped into his mind. He reduced the maximum cross section by giving the fuselage a "wasp waist" where the wings were attached (figure 2.6). The total cross-sectional area of the fuselage and the wings was reduced at their junction to the former area of the fuselage alone. Whitcomb called his discovery the *area rule*—a name that stuck, although, as he noted later, his "rule" had the attributes of a theory: it explained "all of those pieces of information that you've been trying to fit together."[30]

Although the detailed reasons for Whitcomb's discovery are obscure, perhaps the best explanation is one given years later by an aeronautical engineer: Whitcomb was "a guy who just has a sense of intuition about these kinds of aerodynamic problems. He sort of feels what the air wants to do."[31]

Whitcomb's area rule had a lasting effect on the aerodynamic understanding of supersonic flight characteristics. It made possible the immediate redesign of a Convair fighter that was then in trouble because the supersonic drag was much higher than expected, and it guided the future approach to supersonic design. At 34, Whitcomb was awarded the prestigious Collier Trophy by the National Aeronautic Association for "the greatest achievement in aviation in 1955."

Some analysts questioned Whitcomb's originality, seeing the area rule as merely providing "the engineering data that turned [earlier theoretical work] into useful applications."[32] The historian James R. Hansen concedes that the mathematical approach to the supersonic problem taken earlier by the British aerodynamicists G. N. Ward and W. T. Lord could have provided Whitcomb with clues to the area rule; so could a 1947 Caltech doctoral thesis by Wallace D. Hayes. But, as Hansen notes, "perhaps, because they reduced everything mathematically—which involves thinking with symbols—Ward, Lord, and Hayes had failed to *see*, as Whitcomb would, how to bring the physical elements together in a new aerodynamic combination."[33]

Really Seeing

In 1970, astronaut Fred Haise, a few hours into his first space flight, was excited by his astonishing view of the world. He saw New Zealand and Australia laid out below as a full-size map—but no map could be so real

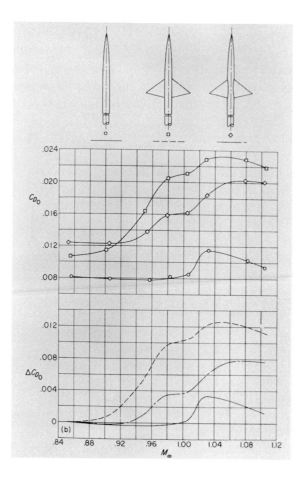

Figure 2.6
The drawings at top illustrate wind-tunnel models of a tapered cylindrical fuselage, the same fuselage with delta wings, and the same fuselage "pinched" in the vicinity of the wings. The three upper curves show absolute drag at Mach numbers from 0.84 to 1.12; the three lower curves show increases of drag over fuselage without wings.[24]

The Mind's Eye

and vivid. Nothing he had ever seen—science fiction movies, flight simula-
tors, or photographs from previous missions—had prepared him for the
staggering experience of reality. He was reminded of the first time he had
seen an elephant, in a zoo. He had seen pictures of the planet just as he had
seen pictures of elephants, but, as he now realized, nothing could have
prepared him for the real thing.[34]

"Seeing a real elephant" is a common experience. A knowledgeable
person who has been familiar with a famous painting or a historic machine,
but has never seen the original, is usually quite surprised in the presence of
the real thing. "I didn't realize it was *that* big" and "I didn't imagine that it
looked like *that*" are standard reactions. There is no adequate substitute or
surrogate for reality.

Robert Pirsig puts intimate firsthand knowledge at the center of his book
Zen and the Art of Motorcycle Maintenance. In a memorable passage, he
tells us what to do when we're stumped, when we have diagnosed "the"
trouble and then found we were wrong:

> . . . just *stare* at the machine. There is nothing wrong with that. Just live with
> it for a while. Watch it the way you watch a line when fishing and before long,
> as sure as you live, you'll get a little nibble, a little fact asking in a timid,
> humble way if you're interested in it. That's the way the world keeps on
> happening. Be interested in it.[35]

I was fortunate to learn early that an engineer's intelligent first response
to a problem that a worker brings in from the field is "Let's go see." It is
not enough to sit at one's desk and listen to an explanation of a difficulty.
Nor should the engineer refer immediately to drawings or specifications to
see what the authorities say. The engineer and the worker must go together
to the site of the difficulty if they expect to see the problem in the same
light. There and only there can the complexities of the real world, the stuff
that drawings and formulas ignore, be appreciated.[36]

Creativity
The notion of "creativity" in engineering, which blossomed in the 1950s,
was intended to describe the ability of some minds to synthesize new ideas

from a combination of past and present experience or from elements experienced separately. The notion was trivialized by a fad, best described as a creativity craze, that swept through US engineering schools in the late 1950s, in the post-Sputnik period of public hysteria when the Russians were ahead of the United States in space and (we were encouraged to think) in military hardware.

In many engineering schools, new techniques to encourage creativity were expected to yield bright ideas that would result in inventions. If Americans could just become more creative than the Russians, it was believed, the United States could forge ahead of the USSR in the fields that counted: space and war.

Courses in creativity were started in engineering schools and in the armed forces willy-nilly. Psychologists staked out a claim in the new field, but nearly anybody could play the game. One of the most successful gurus of creativity was Alex Osborn, a co-founder of the Batten, Barton, Durstine, and Osborn advertising agency. Osborn's chief contribution to the creativity craze was the technique of "brainstorming." A group of people was assembled, the leader explained the occasion for which ideas were wanted, and a secretary wrote down all the ideas that occurred to members of the group. The group was admonished to suspend all inhibitions, and all ideas were recorded—the good, the bad, and the silly. The immediate aim of brainstorming in creativity classes was to develop "ideational fluency"—that is, to produce lots of ideas quickly. The gurus assumed that if enough ideas could be brainstormed, some of them would *have* to be good ideas.[37]

In the early 1960s the National Science Foundation sponsored several national conferences on "scientific" creativity, but by that time the fad had nearly run its course. As interest in engineering design faded in most engineering schools, creativity was put on a back burner.[38]

More important to a designer than a set of techniques (empty of content) to induce creativity are a knowledge of current practice and products and a growing stock of firsthand knowledge and insights gained through critical field observation of engineering projects and industrial plants. In the 1950s, engineering schools still provided many opportunities to gather such knowledge. It is ironic that the radical change in curricula that occurred during the 1950s eliminated those activities that put students in touch with the authentic world of engineering.[39]

Nonvisual, Nonverbal Knowledge
The machines and structures designed by engineers could not be built if sensual knowledge in shop and field did not range far beyond its visual component. In erecting a machine, such as a large steam turbine-driven electrical generator, not only visual but also tactile and muscular knowledge are incorporated into the machine by the mechanics and others who use tools and skills and judgment to give life to the visions of engineers.

Those workers—machinists, millwrights, carpenters, welders, tinsmiths, electricians, riggers, and all the rest—supply all made things with a crucial component that the engineer can never fully specify. Their work involves the laying on of knowing hands. It is sad that engineering schools teach contempt, not admiration, for those hands. It apparently seems more important to maintain the engineers' status and control than to acknowledge the indispensability of the hands and to welcome workers' warnings and insights when (or, preferably, before) things go wrong.[40]

The historical significance of workers' knowledge had hardly been noticed until the English economic historian John R. Harris, in 1971, connected it to the technological lead that Great Britain held over the Continent during the Industrial Revolution. In the seventeenth century, Britain had converted to coal as an industrial fuel while the Continent had continued to use wood. In the glass industry the shift to coal entailed putting molten glass in covered pots to prevent contamination by coal smoke. In the cast steel industry new crucibles, with different mixtures of clay, charcoal, and shards of shattered crucibles, were needed to contain the molten metal at the new, higher furnace temperatures, and stokers learned the subtleties of controlling the temperature of the coal fire beneath the crucibles.[41]

The list of changes of techniques and apparatus is very long, but these changes are unappreciated because many (probably most) of them were made by workers—"especially the senior skilled workmen," Harris notes—rather than by the owners or the supervisors of the works. By 1710 (50 years before the date most historians recognize as the beginning of the Industrial Revolution), workers' growing knowledge of the techniques of coal fuel technology had already given Britain a commanding industrial lead over France and other Continental countries—a lead to which the French responded with a century-long attempt to duplicate British steel. "Those who warned [the French government] that it was men not technical recipes that had to be obtained in order to succeed with the new processes were generally right," says Harris. In 1752, a frustrated French official observed that "the arts never pass by writing from one country to another, eye and practice alone can train men in these activities."[42]

The tacit knowledge and the skills of workers may not have been the determining factors in Britain's leading role in the Industrial Revolution, but they were essential components of it. Today, similarly, the knowledge and skills of workers—sensual nonverbal knowledge and subtle acts of judgment—are crucial to successful industrial production. Yet the engineering profession makes little effort to give credit to skilled and knowledgeable workers or to learn from them.

Few design engineers are expert machinists or welders or millwrights or riggers, but young engineers can learn important lessons about the latent possibilities and limits of craft knowledge and skills if they will but watch experienced workers in their expert, unselfconscious performances. And ask them questions. A designer who spends time intelligently observing field and shop work can expect to learn how to improve the construction of a project and to avoid the surprises that too often result from an engineer's ignorance of the nature of manual skills.[43]

3 Origins of Modern Engineering

The technology of the middle ages [was] the technology of the artisan. [In the Renaissance] the artisan has been split up into his components, the worker and the engineer.

José Ortega y Gasset, 1940[1]

In the later 1480s the new technology became more diffused. For nearly a century thereafter Italian engineers scattered over Europe, from Madrid to Moscow and back to Britain, monopolizing the best jobs, creating wonderful new machines, building palaces and fortifications, and helping to bankrupt every government which hired them. To tax-paying natives they were a plague of locusts, but rulers in the sixteenth century considered them indispensable.

Lynn White, Jr., 1972[2]

The Continuity of Engineering
The strength of engineering lies in the depth of its foundations. The hundred generations of conscientious artisans who built those foundations preserved the technical knowledge they had learned from their forebears, refined it, added to it, and passed it on to posterity. Although we pay attention chiefly to new and showy technical marvels, at least 80 percent of engineers work with technologies that have been around for decades or even centuries. The engineered infrastructure that enables modern society to hold together— buildings, heating and lighting systems, water supply, roads and bridges, self-powered vehicles of all kinds, telephones, and much more—rests on technical knowledge that has been accumulating for as long as *homo faber* (man the maker) has been at work.[3] The unexpected sophistication of bronze water pumps used in Roman mines 1750 years ago (see figures 3.1–3.4) attests to the very early appearance of ideas and techniques that are still widely used.

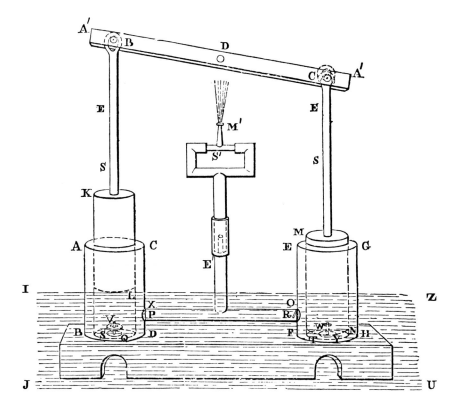

Figure 3.1
Ancient force pump with adjustable nozzle: an 1851 drawing of Hero of Alexandria's
pump (first century A.D.).[f25]

Figure 3.2
Bronze Roman force pump and adjustable nozzle (third century A.D.) found in a mine at Valverde del Camino, Spain.[26]

Figure 3.3
Detail of adjustable nozzle in figure 3.2.

Even the practice of engineering has a continuity that surprises the engineer who looks only at the exciting changes that have come to the engineering office during the past 20 years: from the slide rule to the computer, from the drawing board to the computerized "workstation," from a shelf of reference books to an on-line data base, from sending out charts and tables to be photostatted to the use of computer printers and Xerox machines at every turn. Yet the underlying essence of engineering is unchanged. Preliminary planning is still a process that employs whatever information is readily available. Long before their import can be fully known,

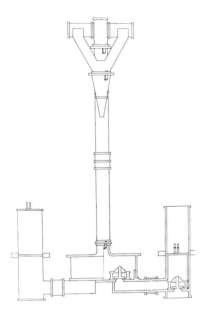

Figure 3.4
A measured drawing of the pump at Valverde del Camino made in 1968.[27]

alternatives are considered and the "big" decisions are made—decisions that determine a project's characteristics for its entire duration.

In the course of developing an engineering design, the designer uses a body of engineering knowledge of great depth and breadth. Consider for a moment an example that suggests both the nature and the often remote origins of that engineering knowledge.

An auto engine is an everyday machine whose existence 500 years ago is impossible to imagine. Yet except for the electrical components—the ignition coil and the spark plugs—nearly all of its elements were known when Leonardo was alive (1452–1519). The engine is composed of cylinders and pistons, a crankshaft, conical valves, cams, gears, bearings, chains, belts, and other mechanical components. The repertoire of mechanical elements was astonishingly close to completion when Leonardo was filling his notebooks with drawings of them. Some of the components, such as cylinders and pistons, date as far back as the first century A.D.[4]

In designing a new machine, an engineer employs familiar components, often in rearranged configurations and occasionally in radically modified ones. In the fifteenth century, an apprentice engineer might have assembled his repertoire of familiar components through careful observation of town clocks, of pumps used to supply fountains, of flour mills, and of hoists and other machines for handling heavy construction materials. He might also, had he proper connections, have studied the drawings in the dozens of engineers' notebooks compiled during that century.

Renaissance Engineers' Notebooks

Today, the most widely known of the Renaissance notebooks are those of Leonardo da Vinci. The reproduction of his sketches and explanatory notes (the latter written in a distinctive mirror script) in countless scholarly and popular publications has built up the impression that much of modern technology can be traced to Leonardo. Because sketches from Leonardo's notebooks are attractive and readily available, they are often used by museum curators as backdrops for displays of modern machines or structures. The extensive researches of Gustina Scaglia, Ladislao Reti, Bert Hall, Frank Prager, Bertrand Gille, and others have now called attention to earlier engineers' notebooks and have traced the origins of many specific drawings in Leonardo's notebooks to earlier notebooks.[5]

Scaglia, who has examined with uncommon perception the manuscript drawings and surviving machines of Italian engineers (from the early fifteenth century) to the end of that century, has identified a network of artist-engineers whose notebooks combined originality and copywork. In correlating the drawings of ten engineers' notebooks to show the exchange of information among engineers, Scaglia noted the repetition of cranes and other machines for lifting and moving the heavy materials used in monumental buildings. She identified the 23 crane types appearing most often in the notebooks. Of these 23 types, each notebook, on average, contained 13.[6]

Lifting devices were but one of several preoccupations that are evident in those notebooks. Another was military machines. Hero of Alexandria, active during the first century A.D., designed missile-throwing devices and included them in his treatise on mechanics in a chapter entitled "Engines of War." The Greek geometer Archimedes and the Roman architect-engineer Vitruvius also designed war machines. Other subjects of wide interest were the raising of water for public and private fountains and the design of water-powered flour mills, sawmills, and other industrial units.[7]

Thus, even before the publication of the first printed book of mechanical technology—the 1472 military treatise of Roberto Valturio—an impressive stock of mechanical knowledge, predominantly nonverbal, had been accumulated in engineers' notebooks. That knowledge was readily portable across cultural, linguistic, and temporal barriers because it was pictorial, requiring few words to explain.

To Design a Fortress

Every Renaissance engineer was engaged at one time or another in designing war machines and fortresses. In 1495, the kingdom of Naples had been overrun by Charles VIII of France, whose army had swept triumphantly down the western coast of Italy with 40 cannon. The new bronze cannon, light enough to be moved over long distances and yet capable of firing iron balls heavy enough to pulverize the masonry walls of any fortification that resisted them, suddenly made obsolete nearly all the castles and walled cities of the ancient and the medieval world. At Monte San Giovanni, a frontier stronghold that had once withstood a siege of seven years was breached and overrun by the French in a single day.[8]

The response of Italian engineers to the painful surprise of the new French cannon was at once imaginative and practical. Accepting the unexpected vulnerability of high masonry walls, they designed fortresses with a very low profile. Wide ditches, on the order of 30 feet deep, were dug around a fortress and lined with massive masonry walls. The keys to impregnability of a fortress were the bastions, the protruding but protected gun platforms at each corner of the perimeter of a fortress. The bastions commanded every square foot of the ditch. Because an attacking force would have to cross the ditch once a wall was breached by cannon, the guns in the bastions were arrayed to rake the ditches with deadly fire.

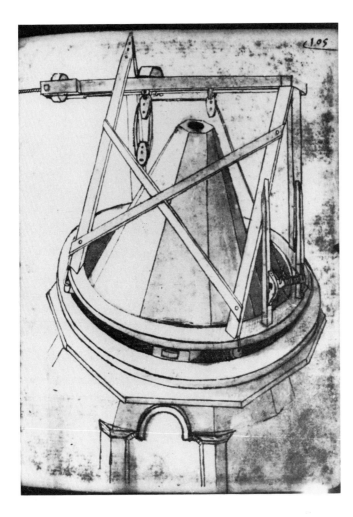

Figure 3.5
A crane designed by Filippo Brunelleschi and used in the construction of the dome and lantern of the Florence Cathedral (1418–c. 1440) is shown in this drawing by Buonnaccorso Ghiberti.[28]

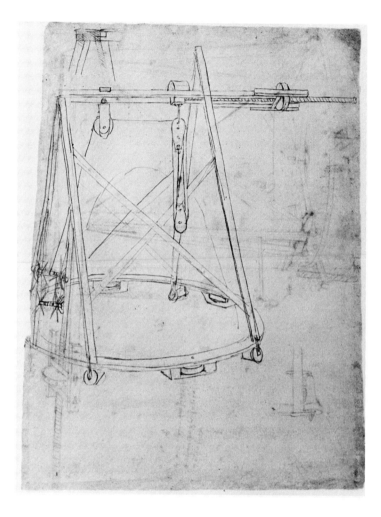

Figure 3.6
Ghiberti's drawing (figure 3.5) was copied (c. 1490?) by Leonardo da Vinci into his
own notebook.[129]

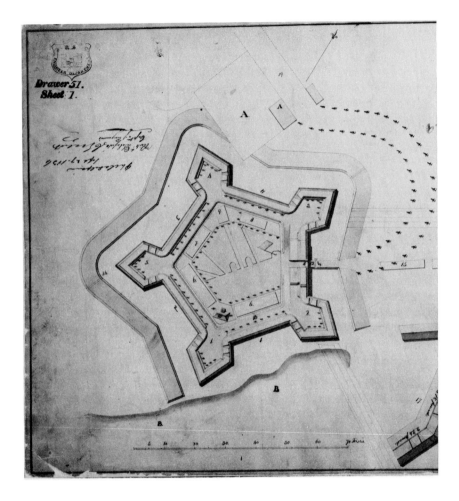

Figure 3.7
Fort McHenry, Baltimore, 1803. This "classic" pentagonal fortress is of the type made
famous by the Duke of Savoy's engineer Francesco Paciotto da Urbino (1504–1576).[30]
The fort was not much altered before the 1814 British bombardment of shells and
rockets, which inspired "The Star-Spangled Banner."

So successful were the newly designed fortresses that Italian design doctrine spread throughout Europe and the New World, persisting without fundamental change for 400 years, until near the end of the nineteenth century.[9]

To create a fortress, the designer laid out on paper a regular polygon with from four to ten sides. A bastion was inserted at each corner. Working out the details of the vast structures—many fortresses surrounding garrison towns were several miles in circumference—became an absorbing problem. The designer had to choose the configuration and the dimensions of walls, the ditches, the gun emplacements, and the lines of fire. All of the design was focused on outwitting and arresting a determined attacking force. The invention and development of the bastion, the glacis, the inner way, the scarp, the counterscarp, and dozens of other elements of the new system of fortification engaged the best efforts of a band of Italian architect-artist-engineers that included such brilliant men as Francesco di Giorgio Martini, Leonardo, and Michelangelo, the first of whom had witnessed the French capture of the city of Naples. By 1550, fortress design had accumulated so extensive and specialized a body of knowledge that a group of able and aggressive military engineers claimed a monopoly on military structures, guns, and machines.[10] For more than 200 years the esoteric arts of fortress design were at the leading edge of high technology, and enormous quantities of capital and labor were expended to build, destroy, and rebuild the massive structures.

The intricate geometric figures that engineers drew on paper led the mathematically minded to the conviction that there must be an ideal form for a fortress. The notion of an ideal form led to a belief in the superiority of a fortress laid out on flat ground, without regard for fortuitous geographic features, such as hills and ravines, the protective qualities of which had for centuries dictated the location of castles. Some engineers rejected the notion of an ideal fortress. Old campaigners, such as Giovanbattista Belluci, argued that a designer should plan with a particular site in mind rather than strive for an unattainable ideal. Younger engineers, such as Galeazzo Alghisi, persisted nonetheless, and soon carried their doctrine across the Alps to the Low Countries. Both camps agreed that, whatever the site, the territory surrounding a fortress must be stripped of any structures or natural cover that might give aid to an attacking force. Cadets at West Point in 1860 were still being taught the precept, laid down 300 years earlier, that "everything on the exterior [of a fortress], within cannon range, which might afford shelter to the enemy, must be removed." A French general of the seventeenth century had observed, more succinctly, that "suburbs are fatal to fortresses."[11]

The Secret of Design

The design of Italian fortresses illustrates a fundamental tenet of engineering—one so right and proper, so self-evident (once it has been pointed out), and now so ingrained that it has become axiomatic: Whatever an engineer may be called upon to design, he or she knows that in order for the plans to be effective the system being planned must be *predictable and controllable*. The designer first defines the boundaries of the system (often involving highly arbitrary judgments), just as the fortress designer set his boundaries at the limits of cannon range. Then the permissible inputs to the system and the permissible outputs from the system are carefully determined. Nothing may cross the boundaries unobserved or unaccounted for. There is no place in an ideal engineering system for unpredictable actions, either by machines or by people. Thus the assumptions and essential procedures of the fortress designer—the fortress mentality—fit naturally the needs of modern engineering design.

Engineers and their Patrons

The Italian Renaissance also illustrates with great clarity the operation of patronage (an inescapable condition of engineering in all ages), which had already reached a mature form. Architects, artists, and engineers constituted a class of creative workmen whose stock in trade was ideas and techniques. These men were needed and respected within the society but were neither economically nor politically independent. Because the common denominator of all patronage was power, both economic and political, engineers and their colleagues needed patrons. The patron might be a high officer of the Church, a powerful nobleman, or a public official. He found money to pay the engineers and to supply them with the people, money, and materials required to build and operate the things the engineers proposed and designed. The patron also provided economic and political legitimacy for the engineers' work.[12]

Patronage is no less important today than it was in the Renaissance. Because an engineer's work is always at least one step removed from the marketplace (he may design an engine, but he is neither its owner nor its seller), he has the same needs of money and legitimacy that his Renaissance forebears had. Today those needs are usually met by government bodies or by corporations.

A famous letter written by Leonardo in 1480, when he was 28 years old, illustrates on a personal level the situation of a young engineer—in the Renaissance or in the twentieth century. Leonardo, who needed an income, was applying to Ludovico Sforza, a leading nobleman of Milan, for a job. In nine brief paragraphs he described his abilities as a military engineer and a designer of siege apparatus, including "bridges of a sort extremely light and strong," "covered chariots, safe and unattackable," portable ladders, mortars, cannon, mines, and catapults. "And, in short," he wrote, "I can contrive various and endless means of offense and defense." In times of peace, he went on, he could be an architect, sculptor, and "can do in painting whatever may be done, as well as any other, be he whom he may be."[13]

Today's engineer, writing a letter of application to a corporation or a government agency, is in the same subordinate position as Leonardo, and for the same reasons. He may have unlimited abilities, ideas, and enthusiasm, but he must persuade a patron to supply the means and the social standing that will give legitimacy to his work of designing ways to change the world.

Toward the end of the Renaissance, the wide-ranging artist-engineers were gradually replaced by specialists who could concentrate their energies on a more restricted repertoire of interests and abilities. The first specialty was fortress engineering; successive subdivision of fields produced artillerists and engineers of siegecraft. In European navies, ship designers became naval engineers. In civilian pursuits, engineers specialized in bridges and roads and in waterworks and canals. In each of the many branches of engineering that grew out of the Renaissance, the ability to use and convey visual information was (as it continues to be) a requirement. The affinity of engineering for art has been masked by the rise of the physical sciences, but the successful practice of engineering will always be shaped by the disciplines of art.

Whereas modern engineering had its origins in Italy, it was the French enthusiasm for building bureaucratic organizations that gave both military

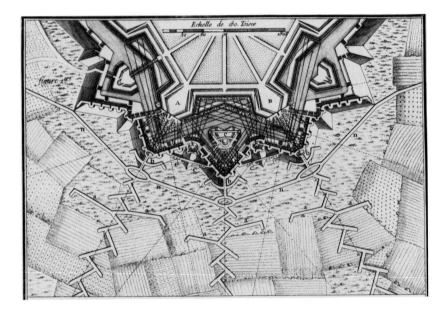

Figure 3.8
Siege warfare of the eighteenth century. A generalized approach to attacking a besieged fortress is shown in this plan view. Starting beyond cannon range, a series of zig-zag trenches was dug by the besieging force. Artillery and other equipment were moved in the trenches to within point-blank range of the fortress. By concentrating artillery fire on a short section of the fortress wall, the wall might be razed, whereupon a large force of trained men might be successful in overrunning the fortress. The drawing shows lines of fire of attacking artillery.[31]

and civil engineering their permanent shapes. The cleverly designed fortresses of the Italian military engineers had neutralized the new artillery, so defensive forces seemed secure for a while. In the seventeenth century, however, the military engineers of King Louis XIV, led by Sebastien le Prestre de Vauban, swung the balance in the other direction by developing siegecraft.

Vauban, one of France's best-remembered heroes, devised a system employing an intricate network of trenches that allowed men and cannon to advance to strong offensive positions in front of a hostile fortress. When that force was in place, its cannon might breach a wall and allow an elite force of sappers, trained by engineer officers, to overrun the fortress. Vauban declared that an impregnable fortress could not be built. He proved his boast in the fifty sieges that he directed in a long and brilliant career. Yet he also devised a new system of fortification, a modification of the Italian system, that was accepted as definitive by many generations of military men.[14]

Schools of Engineering

During Vauban's lifetime (he died in 1707) there were no formal schools for engineers. In the hundred years after his death, however, a system of engineering education was developed in France. In the nineteenth century that system spread throughout most of the rest of the world. Indeed, the philosophy that underlies engineering education today is the same philosophy that informed the organization of the French schools of the eighteenth century. Engineering students in twentieth-century America, like their French counterparts of 200 years ago, must prove their mathematical abilities, their knowledge of the physical sciences, and their willingness to spend long hours on carefully prescribed numerical problems.

Artillery schools were the first to provide what came to be known as a mathematical education, which was and is the fundamental theoretical constituent of a technical or scientific curriculum. By 1720, artillery schools had been opened in several French garrison towns, where cadets were given a grounding in algebra, geometry, trigonometry, and engineering mechanics. The first school for military engineers (as distinguished from artillerists) was established in 1749 in Mézières, a fortified town near the northeastern border; though not large, this school had a crucial influence on the famous Ecole Polytechnique, which was established during the French Revolution.[15]

The quasi-military Corps des Ponts et Chaussées (Corps of Bridges and Roads), responsible to the king for designing and supervising the construction of civil works, opened a school in Paris in 1775. All candidates for the Corps spent three years in the Ecole des Ponts et Chaussées, whose curriculum included the same sequence of mathematics and mechanics that was followed in the artillery schools.[16]

In the school at Mézières, a young instructor of physics, Gaspard Monge, originated descriptive geometry, which provided a powerful graphical approach to the solution of many problems in three-dimensional space. By the time of the revolution, Monge had moved on to Paris. A member of the ruling Convention, he was one of a small group of savants who marshalled the technical resources of France to supply the military needs of the revolutionary forces. In 1794, he guided the establishment of the Ecole Polytechnique in Paris. That school became the hub from which spokes radiated to "schools of application," comprising artillery, military and naval engineering, bridges and roads, and mines. The Ecole Polytechnique, a two-year school whose students were chosen by stiff competitive examinations, provided the mathematical core of an engineering education. Its graduates went on to the schools of application to complete their education as specialized engineers. From that time forward, engineering graduates in France enjoyed high political status. The schools they attended became known as "les grandes écoles."[17]

Engineering education in the United States followed closely the precedents set in les grandes écoles, particularly in adopting the central core of mathematical studies. The US Military Academy, when it was reorganized in 1817 by Colonel Sylvanus Thayer, adopted the curriculum of Ecole Polytechnique and some of its textbooks. The Rensselaer Polytechnic Institute also owes its philosophy directly to the Ecole Polytechnique, and almost all the American engineering schools followed the same pathway. There have been some changes, but American engineering schools have remained close to their original spirit and substance.[18]

4 The Tools of Visualization

The tool-maker wants not a verbal description of the thing he is asked
to make but a careful picture of it. . . . without pictures most of our
modern highly developed technology could not exist. Without them
we could have neither the tools we require nor the data about which
we think.

William M. Ivins, Jr., 1953[1]

Nor must we forget that the invention of that method of rigorous description of natural reality, which was the work of the great artists of
the fifteenth century, has the same importance for the descriptive
sciences . . . as did the invention of the telescope or the microscope
in the seventeenth century.

Paolo Rossi, 1962[2]

Great advances in our abilities to convey visual information began during the
Renaissance. A series of fundamental graphic inventions, including printing,
linear perspective, and projective geometry, greatly enhanced the precision
with which a vision in one person's mind might be conveyed by material
means—drawings—across space and time to another person's mind.

Until the fifteenth century, the world of learning was a world of manuscripts. Copyists made copies of manuscripts, and scriptoria existed in which
many scribes made simultaneous copies of a manuscript by "taking dictation"
from a single reader. But copies inevitably incorporated errors and changes,
and scholars knew by experience that their copies of works by ancient authorities were corrupted in the process of transmission. A drawing or a diagram
in a mechanical or mathematical work was subject to even more rapid corrup-
tion than was straight text. A critical reader in a scribal culture was well
advised to assume, with Archbishop James Ussher, that "the ancientest must

needs be the right, as the nearer the Fountain the purer the streams, and that errors sprang up as the ages succeeded."[3]

The art of printing, which in Europe originated around 1450, marked a watershed in the ability to convey information, both verbal and visual. It expanded dramatically the number of readers an author might reach. In the 50 years after the first printed book appeared, 8 or 10 million copies of nearly 40,000 different titles were printed—an average of two new books a day, each issued in an edition of 200 or 250.[4]

Furthermore, the ability of printed books to duplicate exactly in hundreds of copies not only text but also drawings and diagrams was of revolutionary significance in science and technology, where visual information is quite as necessary as verbal. Copper-plate engravings, numerous by the latter part of the sixteenth century, permitted finely detailed drawings to be produced in large numbers. Woodcuts, while not as precise as copper-plate engravings, nevertheless made it possible to multiply the work of a draftsman by hundreds.

The way we think about printed and engraved technical drawings was changed by an insight of William M. Ivins, Jr., curator of prints at the Metropolitan Museum of Art in New York. In his book *Prints and Visual Communications,* published in 1953, Ivins argued that the ability to make "exactly repeatable pictorial statements" was a crucial ingredient in the rise of modern technology and science.[5]

To appreciate the power of Ivins' insight, consider the thousands of different drawings and charts that a New York engineering and construction firm prepared for the Tarbela dam project discussed in chapter 1. Without the means to make many copies of each drawing and each chart for the various contractors, subcontractors, and suppliers of components, the construction of that dam in the way planned would have been unthinkable. Or consider the fact that a modern airplane requires thousands of different drawings, all of which are copied in quantities from dozens to hundreds. The British VC-10 passenger jet required 50,000 different production drawings in addition to many hundreds of design drawings.[6]

How exact the "exactly repeatable pictorial statements" must be was implied but not specified by Ivins. To most of us, perhaps, the power of

Ivins' insight is obvious. However, because some readers may be interested to see how minor were some of the copyists' corrupting changes, several examples of fifteenth-century engineers' drawings are presented in an appendix at the end of this chapter.

Pictorial Perspective

The most significant graphic invention of the Renaissance was *pictorial perspective* (also called *linear perspective*), which produced a qualitative change in the ease with which a visual image in one mind could be conveyed to another mind. Since their invention in the fifteenth century, perspective drawings have provided a uniform convention for pictorial representations of the three-dimensional objects. Such drawings can be interpreted with little effort by most viewers.[7]

The differences between technical drawings before and after linear perspective was commonplace are epitomized in two drawings of an up-and-down sawmill—the first (figure 4.1) drawn around 1230 in the notebook of Villard de Honnecourt, the second (figure 4.2) published in 1578 by Jacques Besson. Villard's drawing is ambiguous, although it would not be difficult for the mind's eye of a reader who had seen an up-and-down sawmill to move the elements into their proper places. On the other hand, Besson's drawing tells a reader what the sawmill looks like and how it operates. (An operator supplies the power to swing the pendulum back and forth. Threaded collars around the oscillating pendulum shaft move the upper ends of the lazy tongs closer together and farther apart, thus lowering and raising the "sash frame" that holds the saw blades.) The inclusion of the operator in this drawing shows the relative size of the machine. The upper frame supporting the saw and its up-and-down mechanism is skewed with respect to the supporting posts, which extend to the floor; however, this skewing makes the details of the saw and its sash frame clearer than would a geometrically consistent view.

The fundamental principle of pictorial perspective is illustrated in figure 4.3. An artist looks through a fixed eyepiece and records the outline of his subject on a transparent screen. The eyepiece and the transparent screen locate and define a visual pyramid of light rays converging from base to

The Tools of Visualization

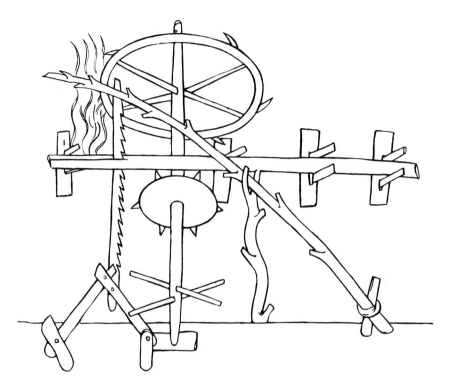

Figure 4.1
A water-driven up-and-down sawmill of the first half of the thirteenth century, from a notebook of Villard de Honnecourt (probably an architect or *maître d'oeuvre*).[f32]

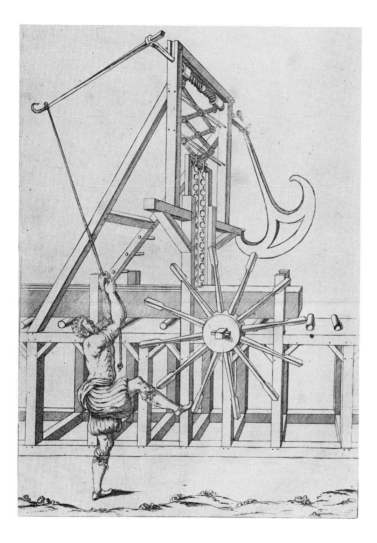

Figure 4.2
Perspective view of manually driven up-and-down sawmill from Jacques Besson's pioneering machine book *Theatre des instrumens mathématiques et mécaniques.*[133]

The Tools of Visualization

Figure 4.3
An artist drawing a portrait in perspective, 1525. The artist looks through a fixed eye-piece at the subject. With a small brush, he paints outlines on the glass surface in the wooden frame before him. His eye is at the apex of the visual pyramid. The difficulty of transferring the painted marks on the glass to a canvas makes the apparatus of figure 4.4 an attractive alternative.[34]

Figure 4.4
Here the artist records directly on a ruled sheet what he sees from a fixed viewpoint through the wooden frame. The rectangular divisions of wire within the frame enable him to draw outlines in the corresponding blocks of the sheet before him, thus eliminating the difficulty of the apparatus in figure 4.3.[35]

apex. The same principle is involved in the apparatus of figure 4.4. Here, the artist looks through the vertical grid and records the outlines of his subject on the ruled sheet under his hands.

The first geometrically constructed perspective drawings of the Renaissance were produced around 1425 by Brunelleschi. Although his drawings have not survived, the rules he followed were codified and published by Leon Battista Alberti about ten years later.[8]

"Alberti's window," as the picture plane (the glass in figure 4.3; the grid in figure 4.4) came to be known in Alberti's system of perspective construction, was promptly adopted by many Italian artists of the fifteenth century, including Leonardo, although some artists continued to use an empirical perspective developed in thirteenth and fourteenth centuries. (The latter complained, with some justification, that a large picture, which will be seen from many different points of view, looks artificial when the artist slavishly follows Alberti's rules, because linear perspective "by the book" depicts what is seen by "a one-eyed observer rooted to the spot."[9])

The Tools of Visualization

Generally speaking, engravers employed linear perspective from the fifteenth century through the nineteenth.[10] The introduction of half-tone printing near the end of the nineteenth century made possible the mass duplication of photographs. The modern camera, a "perspective machine," automatically records a perspective view.[11] Photographs thus continue a 400-year tradition of representing three-dimensional technical objects, albeit with lessened contrast (which results in a loss of sharpness in outline and clarity of details). Drafters still construct perspective drawings when precise information must be conveyed, as in mechanical, botanical, and medical drawings. Such drawings represent a rhetorical version of an object, with certain information emphasized.

The cutaway view and the exploded view were minor graphic inventions of the Renaissance that also clarified pictorial representation. In a technical drawing, details of machine parts hidden by the casing or by other elements are often shown in cutaway views, with as much of the obscuring element removed as is necessary. Figures 4.5 and 4.6 (from a notebook of Mariano Taccola, c. 1430) are precursors of the cutaway view. They simply let the reader look through an obstruction by assuming it to be transparent. In 1556, cutaway views in definitive form were used in Georg Agricola's mining book *De Re Metallica* to illustrate underground operations (figure 4.7). An exploded view, in which the parts of a machine are spread out along a common axis, reveals details of individual parts while also showing the order in which they should be assembled (figures 4.8, 4.9). These views, which help an engineer show reality in rigorous yet imaginative ways, originated in Taccola's notebooks of the early fifteenth century and were perfected by Francesco di Giorgio and Leonardo. They are still useful, especially in operating and repair manuals for intricate machines.

Orthographic Projections
The term *engineering drawing,* as used by an engineer, refers to an orthographic projection, usually one showing three views of its subject. Lines are dropped perpendicularly from points on the object being depicted to the paper on which the projection is to be drawn. A prototype that dates from 1528 is a set of drawings by Albrecht Dürer showing three views of a human

head and three views of a foot (figures 4.10–4.12). With these drawings, Dürer taught artists (and eventually modern engineers) how to describe three-dimensional objects exactly on flat paper.[12] Peter Booker, historian of engineering drawing, sees the head as "crated" (as is the foot) "in an imaginary box which just fits it"—a conceptual crate now readily recognized but not generally employed by draftsmen until the nineteenth century.[13] In Dürer's drawings, the crates were turned through right angles and the projections on the several sides of a crate were recorded on a single sheet of drawing paper.

Gaspard Monge, the eighteenth-century inventor of descriptive geometry, is often cited as having introduced orthographic projection to engineering drawing. However, the nineteenth-century development of three-view orthographic projection was not a definitive method handed down from above but a collaborative effort of teachers, textbook writers, and anonymous draftsmen in Europe and America. Monge's book on descriptive geometry, published in 1795, dealt with orthographic projections, but a reader would find firm guidance for only two views: plan and elevation. Booker found that Monge's book had little influence on drafting practice in Great Britain or the United States,[14] and Yves Deforge, author of a French history of technical drawing, states flatly that "technical drawing is not the child of descriptive geometry."[15]

An early series of machine drawings that exhibit meticulous use of orthographic projection were made by James Watt and his assistant John Southern between 1775 and 1800. Most of these, now preserved in the Birmingham [England] Central Library, show two views of an engine or of a subassembly. When two views were insufficient to describe all the necessary details unambiguously, a third was added.[16] The standard arrangement in the United States and Canada places the top view immediately above the front view; the side view is to the right of the front view. European drawings are arranged differently. Both arrangements can be readily understood by a careful reader.[17]

US patent drawings, which supplement a patent's text, follow Patent Office rules, which are unlike any industrial standards for drafting. To those familiar with conventional engineering drawings, patent drawings have a

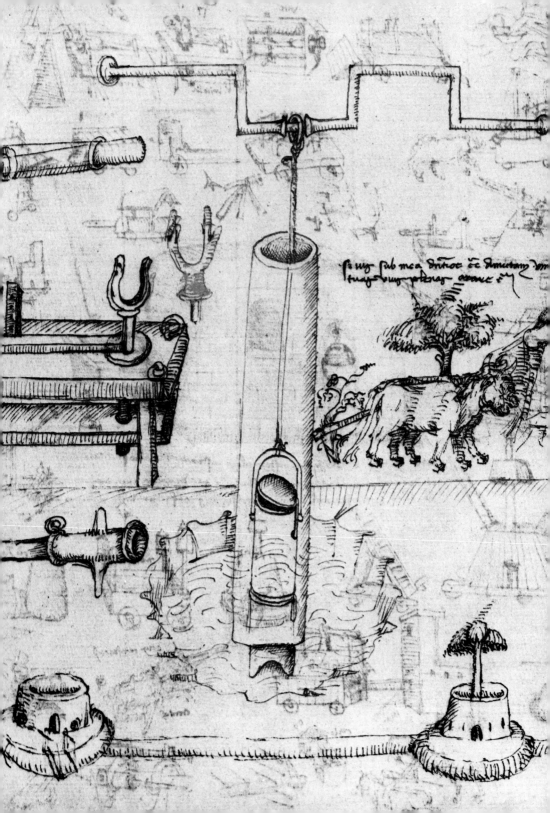

Figure 4.6
"Phantom view" of a syringe piston and rod, from Mariano Taccola's notebook
(ca. 1430).[f36]

Figure 4.5
"Phantom view" of a lift pump, from Mariano Taccola's notebook (ca. 1430).[f36]

The Tools of Visualization

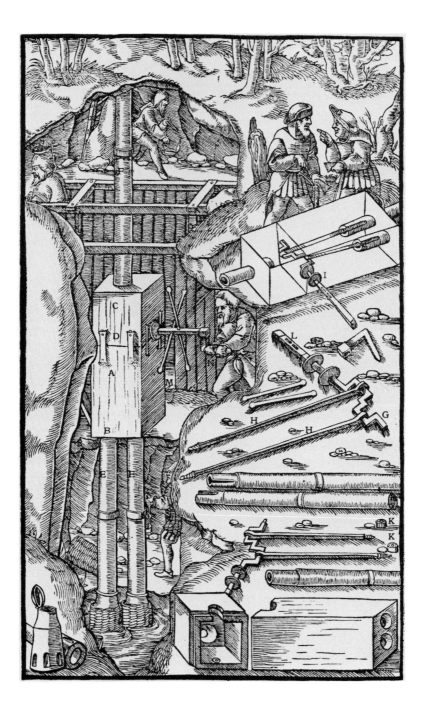

Chapter 4

Figure 4.7
This drawing of a manually operated reciprocating water pump *in situ* also includes a phantom view of the pump case (upper right), an assembly of crankshaft and piston rods (center right), and other disassembled components.[37]

quaint air. Various views, both orthographic and modified perspective, may be scattered on a single sheet. It is presumed that a person "skilled in the art" of the class of devices represented by the patent will be able, by using both text and drawings, to construct and operate the device being patented.[18]

Reading an engineering drawing is a decoding process. Experienced readers know what to look for and pursue the wanted information until they find it or until they are satisfied that it is not on the drawing. Readers of an unfamiliar drawing first build in their mind's eye a three-dimensional picture of the object depicted; they then proceed to whatever details they need to determine the intentions of the drafter. Just as the drawings of any complex object or system require many days to construct, so readers require considerable time to understand them thoroughly. If the reader is someone slated to supervise the building of the system depicted, that person will spend many dozens of hours with a set of drawings to become intimately acquainted with the project, and will return to them hundreds of times.

For those who find such drawings daunting, the process of understanding is not much helped, initially, by the bewildering array of standard symbols designed to reduce drawing time and to avoid ambiguity. The thousands of standard graphic symbols both increase the use of visual jargon and make it easier for initiates to decode drawings. Dozens of different symbols are used to tell welders what kind of welded seam is required. Symbols have been standardized for particular devices, such as pipe fittings, pumps, filters, electric and electronic switches, fuses, terminals, and connectors. Many hundreds of symbols are regularly used. For example, about a hundred symbols are specified for heating, ventilating, and air conditioning, including 20 different symbols for valves (figures 4.13, 4.14).[19]

As an example of an engineering drawing, consider figure 4.15, a mundane, run-of-the-mill three-view orthographic projection of a welded steel

Figure 4.8
Leonardo's ratchet device of c. 1500 is shown assembled (at left) and "exploded" (at right).[138]

"motor support frame." The first impression may be one of confusion, because almost nothing calls attention to itself. There is no evident place to begin. The lettering is uniform and expresses chiefly the jargon of making a steel "weldment" (itself a bit of twentieth-century jargon). Lines that describe shapes are easily confused with lines that carry dimensions. The symbols that show how particular welded joints will be made are unobtrusive yet crucial. This drawing can yield a great deal of information quickly, but only to a reader who knows how to extract it. (The best way to learn how to read drawings, and probably the only fully effective way, is to learn how to make drawings.)

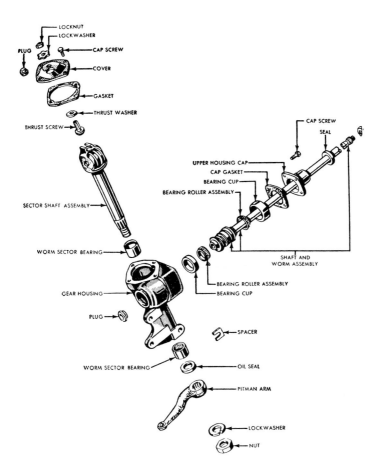

Figure 4.9
Exploded view of a manual steering gear for an automobile, from a repair manual of 1956.[39]

The Tools of Visualization

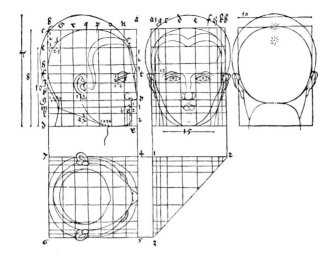

Figure 4.10
Dürer's grid, truncated by a "reflecting line" at 45°, transfers points on the full-face view to the upside-down "plan" view.[40]

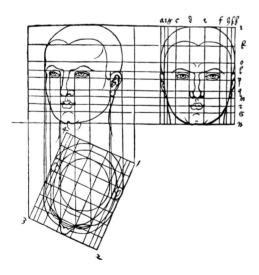
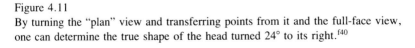

Figure 4.11
By turning the "plan" view and transferring points from it and the full-face view, one can determine the true shape of the head turned 24° to its right.[40]

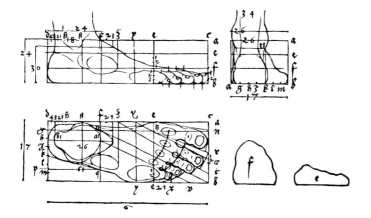

Figure 4.12
The two sections through the foot (lower right) are at f and e of the plan view.[40]

Let us first try to visualize the shape and the size of this motor support frame. We can concentrate briefly on the outline, which is shown in three external views (the fourth view is a section, at upper left). Separating object from ground—that is, seeing the outlines while ignoring dimension lines, numbers, letters, and symbols—enables one to reduce the drawing mentally to the three views shown roughly in figure 4.16. We omit the section for the moment. The upper view is a *top view,* looking down on the motor frame. The view immediately below shows what the orthogonal eye sees when it looks at the near side of the top view; it is called a *side view* or *side elevation.* The view at lower left—an *end view* or *end elevation*—is the result of the orthogonal eye's looking at the near end of the side view.

As we put the pieces together, we notice that the top and side views tell us most of what we want to know. A single steel plate has been bent to a bracket shape. Vertical stiffening triangles (gussets) extend from one end of the bracket to the other. The gussets are shown in the top view as roughly parallel dashed lines on each side of the openings. The dashed lines depict surfaces (the sides of the gussets) hidden beneath the surface of the plate. To learn how big the motor support frame is, we note the overall dimensions: 31½ inches wide by 63⅜ inches long, made up chiefly of steel plates 1 inch thick. The motor is intended to rest on the four rectangular pads attached to the top plate, each 2 inches thick and drilled for holding-down bolts. All

The Tools of Visualization

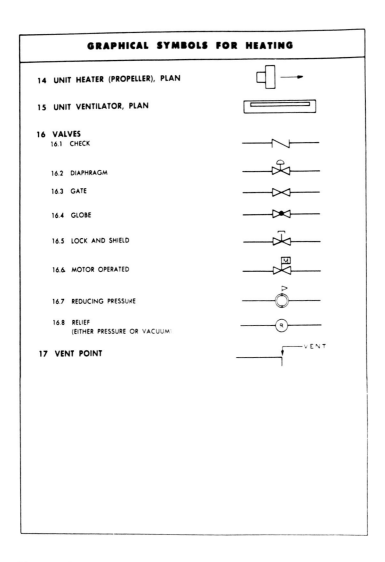

Figure 4.13
ANSI graphical symbols for heating.[41]

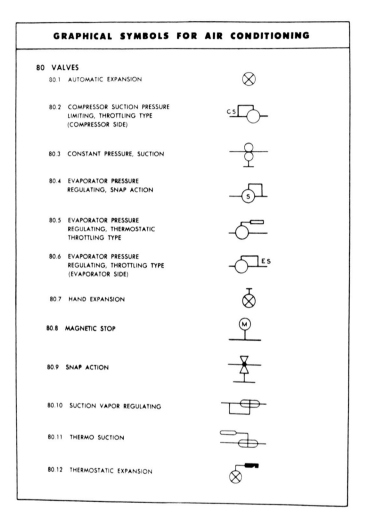

Figure 4.14
ANSI graphical symbols for air conditioning.[41]

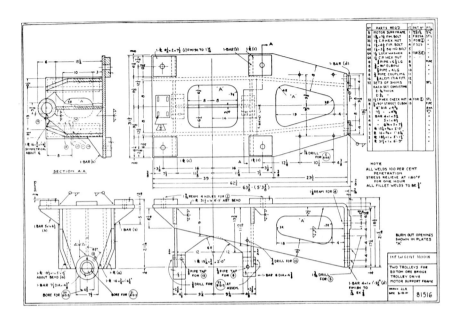

Figure 4.15
Engineering drawing of a "motor support frame."[f42]

three views are employed to gain full information about the pads: their length, width, and thickness are specified in the plan view, the hole size is given in the side view, and instructions for welding are given by symbols in the end view.

The end view clarifies the two rectangles hanging from the left half of the bracket in the side view. They are heavy sleeves, or bearings, bored for a turning shaft not shown but identified by another drawing number.

The fourth view, labeled "Section A-A," tells the reader what would be seen if the assembly were cut apart vertically along a line (broken intermittently by two short dashes) extending across the top view from A to A. The arrows at the ends of the line show the direction in which the orthogonal

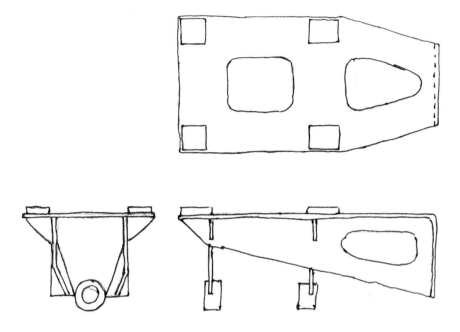

Figure 4.16
Simplified view of a "motor support frame," intended to emphasize the principal elements in figure 4.15.

eye is to look at the section. The sectional view clarifies configurations that can otherwise be deduced only from hidden lines.

Finally, a perspective view of this motor support frame (figure 4.17) seems to furnish information in a more direct way than the orthographic views, yet nearly the reverse is true: in no place on the perspective view is a true shape shown; circles become ellipses, and rectangles have neither right angles nor equal sides. On the other hand, the orthographic drawing shows precise dimensions of shapes and drilled holes. To describe the object in perspective with as little ambiguity as the orthographic drawings provide

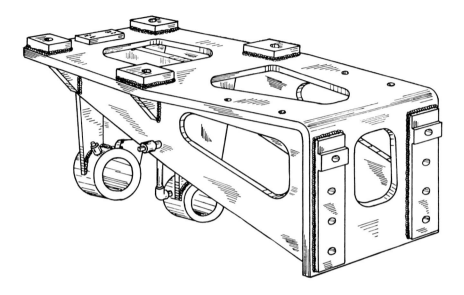

96

Figure 4.17
Perspective view of a "motor support frame." Although the entire object is easier to
visualize than in an orthographic drawing, the maker would have to guess at the exact
shapes of components. An orthographic drawing specifies the actual shapes intended
by the designer.[f42]

would require a number of auxiliary views to clarify hidden and distorted
features.

Making Engineering Drawings
An engineering drawing starts with a series of freehand sketches or roughly
scaled drawings to help visualize how the object being designed will work
and how the finished drawings must be made in order to show a reader
exactly what the designer had in mind. Because design is a complex intellectual process, the designer uses sketches to try out new ideas, to compare
alternatives, and (this is important) to capture fleeting ideas on paper.
    Three kinds of sketches can be identified. The first is the *thinking sketch*.

Leonardo's notebooks contain dozens of such sketches (see, e.g., figure 4.18), and a host of later engineers have used sketches to focus and guide nonverbal thinking. The next is the *prescriptive sketch* (e.g., figure 4.19), which is sometimes scaled and which is made by an engineer to direct a drafter in making a finished drawing. Thomas E. French, whose pioneering college textbook *A Manual of Engineering Drawing* and its successors have been used by 2 million students, declared that "the designer must be able to *sketch* his ideas with a sure hand and clear judgment. In all inventive mechanical thinking, in all preliminary designing, in all explanation and instructions to draftsmen, freehand sketching is the mode of expression. . . . It is the chief engineer's method of design."[20]

The third kind of sketch, produced constantly in exchanges between technical people, is the *talking sketch* (figure 4.20). In the 1980s, a young engineer in the design division of a "mid-size, mid-tech" machine works equipped with computer-assisted design equipment, commented on her initial surprise at the customary mode of communication. Two designers never "just sit down and just talk," she said. "Everybody draws sketches to each other." Some designers would take early versions of their new designs to the fabricating shops in order to solicit suggestions for improvements. In going over a drawing with a welder, one senior designer regularly made "little hand sketches off to one side" in order to clarify complex and possibly confusing parts of the drawing.[21] Kathryn Henderson, a sociologist who is studying the politics of engineering design at first hand in design departments of several industries, remarked on the way talking sketches were made: she "observed designers actually taking the pencil from one another as they talked and drew together on the same sketches."[22] Those talking sketches, spontaneously drawn during discussions with colleagues, will continue to be important in the process of going from vision to artifact. Such sketches make it easier to explain a technical point, because all parties in a discussion share a common graphical setting for the idea being debated.

Using Engineering Drawings
In the eighteenth century, before engineering drawings came into general use in machine-building shops, a client who wanted a machine built described to

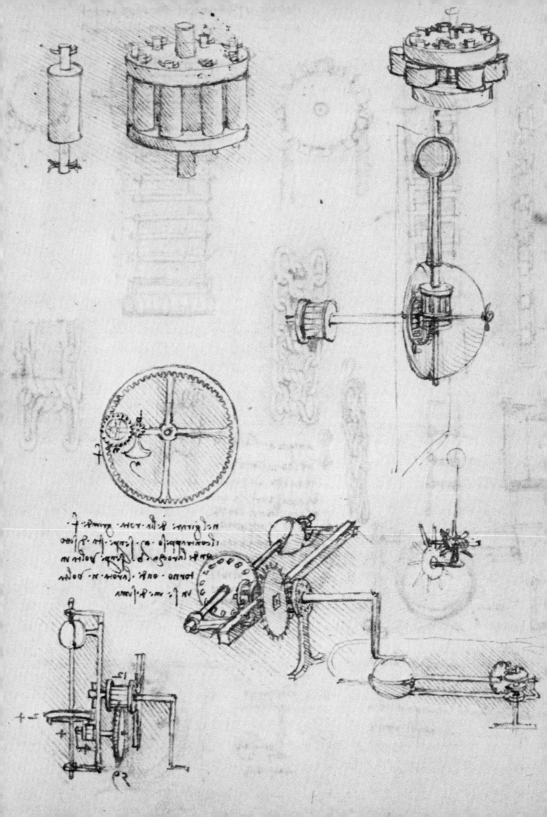

the shop owner or to a foreman what he had in mind. The owner or foreman might consult with his senior workers; the final decisions as to how the desired machine was to be built were then agreed to in a series of negotiations between those who would build the machine and the client who would pay for it.

In the nineteenth century, when engineering drawings were in wide use, the locus of decision shifted from the shop floor to the drafting room. James Watt's drawings, mentioned above, provided precise and uniform information to the Boulton and Watt shops, to outside suppliers of parts (such as cylinders) that were not made by Boulton and Watt, and to the erectors who would arrange locally to build the engine house and to assemble the engine and prepare it for operation.

In 1836, in the Paris shops of the important French machine building firm of Desrone et Cail, the foremen were responsible for deciding how the work should be done. Their authority and responsibility were far-reaching and included acting as inside contractors. They hired their own workers, arranged for materials and tools not supplied by the owners, and sold to the owners the products that they and their workers had designed and built.[23] In that same year, Jules Cesar Houel, a recent graduate of the Ecole des arts et métiers at Chalons-sur Marne, was hired by Desrone et Cail. Houel, one of the first technical school graduates to work for the firm, started as a worker on the shop floor and soon became a foreman supervising a gang of workers. Following the customs of the shop, he also became an inside contractor and negotiated with the firm for the work that he and his crew produced. In 1840, after he had been with the firm for 4 years, he was promoted to a new position as the first technical director of Desrone et Cail.

Figure 4.18
Thinking sketches by Leonardo da Vinci, c. 1500. Engineers use thinking sketches to clarify visions in their minds' eyes. This page of Leonardo's notebooks reflects two different trains of thought. The three sketches at the top are studies to reduce friction in (lantern) gear teeth. Other sketches probe ways to arrange epicyclic gear trains to give proper movement to the moon (spheres in lower views) in an astronomical clock.[43]

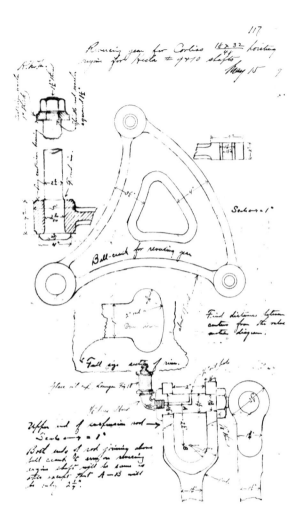

Figure 4.19
A prescriptive sketch of steam engine parts, 1889. Made by Erasmus D. Leavitt
(1836–1902), a prominent designer of very large steam engines, this sketch of re-
versing gear parts for a Corliss engine to be installed in the Calumet and Hecla mines
on the Upper Peninsula of Michigan was used by the draftsmen who made the working
drawings for the construction of the engine. This is a press (office) copy of Leavitt's
original sketch.[44]

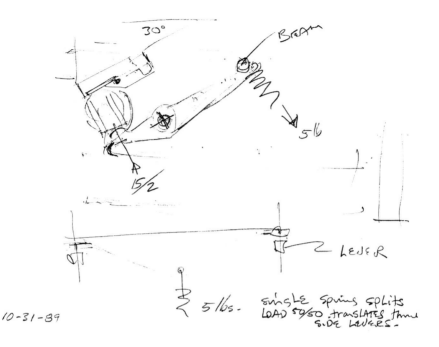

Figure 4.20
A talking sketch of components of a computer printer. This sketch evolved during a
discussion between E. Hubbard Yonkers and his client as the designer explained one
possible way to provide a paper nip force equally distributed across a relatively slender
roller within a minimum vertical height by means of lever arms which straddle the
thermal print head/roller assembly.

Within a relatively short time, Houel reorganized the shops, transferring the
power to make nearly all decisions from the foremen to the drafting room.
He supplied complete engineering drawings of the desired machines and
required the foremen to have the work done exactly as specified on the
drawings. It appears that the foremen continued for a while as inside contractors, negotiating with workers and with Houel, but the drawings changed
radically the balance of power between managers and workers. In 1849,

The Tools of Visualization

Houel's innovation in the control of skilled workers won him a place in the *Legion d'honneur.*[24]

The transformation that occurred in most European and American shops in the nineteenth century was from a world in which engineers negotiated with workers who had traditionally used the judgment of their trades to a world in which matters of judgment were settled by pencils on paper in drafting rooms remote from the shop floor. The removal of all discretionary power was neither sudden nor uncontested, but within a few decades the center of authority in engineering work was clearly located in engineering drawings.

Study Models

Models, a second type of graphic device, have served two chief purposes. One is to supply information to those who are involved with building or operating the full-size structure or machine that the model depicts. The other purpose, as we shall see in the next chapter, is to teach ideas and principles and to acquaint observers with unfamiliar structures and machines. In both cases, models provide tactile as well as visual information. Most of us find a model much easier to read than a drawing.

In fourteenth-century England, Hugh Herland, master carpenter to King Richard II, used models in designing the timber roof structure of Westminster Hall, an arched truss of great complexity that spanned 68 feet of unpillared space. The distribution of stresses in that unprecedented truss has been for many years the subject of a debate, at times acrimonious, among modern analysts. (The argument points up the unavoidable presence of assumptions in engineering analyses and the conflicting conclusions to which different assumptions can lead.[25])

In Renaissance Italy, the selection of an engineer to carry out a project often depended upon competitions that employed models to show how the given objective might be accomplished. In 1418, twenty or more models were submitted by competitors for the job of building the great dome of the Florence cathedral. When Brunelleschi was chosen to plan and supervise the construction, he proceeded to build a larger model (perhaps 15 feet across) in order to work out the details of his proposal to build a masonry dome without false-work—a radical and audacious undertaking. The dome that Brunelleschi built, virtually unchanged to this day, is 140 feet in diameter (the dome of the US Capitol is 85) and is founded on the church's walls more than 100 feet above its floor.[26]

In 1585, when Pope Sixtus V decided to move the Vatican obelisk—a monolith nearly 100 feet high and 330 tons in weight—to its present location in front of St. Peter's, he appointed a commission to review the several hundred proposals put forward by contenders for the job. The successful contender, Domenico Fontana, used an elaborate model to demonstrate how he intended to lift and move the unwieldy monolith. His 2-foot model obelisk was made of lead; the timbers and rigging, including ropes and pulleys, were proportioned accordingly (figure 4.21).[27]

During the reign of Louis XIV in France, models of forts and fortified towns, both French and foreign, were built for army officers to study so that they might become familiar with the possibilities of defending and besieging the forts. (See figure 4.22.) More than a hundred of those models, called *plans-relief,* survive in the attic of the military museum of Les Invalides in Paris. The models exhibit exquisite detail, down to the windows, dormers, and chimneys of ordinary dwelling houses. They are also very large—at a scale of over 8 feet to a mile (1/600), many of the models are 15 or 20 feet on a side.[28]

In 1692, Sebastien le Preste de Vauban was the chief engineer, under the king's command, in the successful siege and capture of Namur, a fortified town on France's northeastern border. In September 1695, the town was retaken by the Dutch engineer Menno, Baron von Coehorn. Vauban was not present at Namur in 1695, but he was apparently blamed by someone for failure to strengthen the fortifications after the 1692 siege. On October 6, 1695, Vauban sent a note to Le Peletier de Soucy, the king's director of fortifications:

> There is a relief [model] of Namur in the Tuileries. I beg you to oblige me and come see it with me. I want you to touch with your fingers and see with your eyes all the vulnerable points of that place, of which there are several, and at

Chapter 4

Figure 4.21
Models submitted by entrepreneurs competing for the commission to move the Vatican obelisk about 800 feet (in 1585) to its present location near St. Peter's cathedral, then under construction. Of the many schemes proposed, that of Domenico Fontana was chosen. His model, about 2 feet tall, is at upper left.[145]

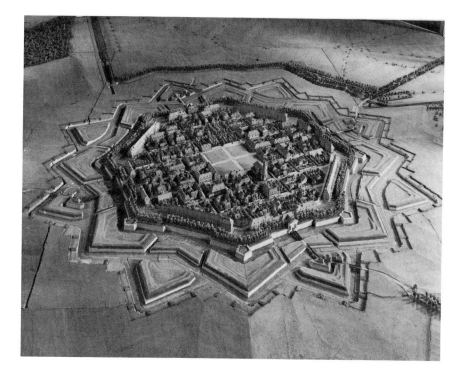

Figure 4.22
Plan relief, c. 1700: a detailed model of the fortified town of Neuf-Brisach, just west of the Rhine in Alsace. The octagonal fortress is about a kilometer across. The model, at a scale of 1:600, is about 7½ × 15 feet overall. Photo: Centre Nationale des Monuments Historique et des Sites.

The Tools of Visualization

the same time to determine how to correct those that are imputed to me; and you will see that they were correctable only by spending large quantities both of time and money. . . . If the King had been able to give us 700,000 écus and 4 or 5 years of time, we might have eliminated some of the weaknesses. To do that, we should have had to rebuild nearly all the original fortifications.[29]

In the nineteenth century, when clipper ships were carrying the premium cargoes of the world, many ship designers and builders used wooden models to develop and describe hull shapes. The models, generally less than 4 feet long, were carved by a designer who was thus enabled to show streamlines more readily in models than in drawings. The models were built up of parallel layers. If the layers were horizontal, the model was called a *lift model*; a *section model* had the parallel layers arranged vertically and transversely, showing the contours of the frames, or ribs, to which the exterior planking was fastened.[30]

In the early 1950s, Admiral Hyman Rickover called for a full-scale model to assist in the building of the *Nautilus,* the first nuclear submarine. According to the historians Richard Hewlett and Francis Duncan, Rickover directed the builder to "mock up in wood and cardboard every pipe, valve, electric panel, and large motor in the reactor and machinery compartments." "The full-scale mock-up," Hewlett and Duncan note, "had a special fascination for Rickover. During visits to Groton he would climb through the simulated compartments . . . [and] study [the] configuration from several angles, insuring enough space for men to maintain and replace equipment at sea and insuring no valve handle extended dangerously into a walkway." Rickover was convinced that the full-scale mock-up provided information that even the most experienced shipbuilder could not gain from drawings or quarter-scale models.[31]

A model can take an observer one step closer to reality than can a drawing. An orthographic projection is the medium *par excellence* for explaining the details of a structure or a machine, and a perspective drawing provides a qualitative sense of the relative proportions of various members. Yet the observer often wants to see around the corner of an object in a perspective drawing, because there (he tells himself) lies the missing piece of the puzzle of understanding. A model permits the observer not merely to look around that corner but to walk around it, to look down on and up at the object, and to receive tactile clues that help him make sense of the object.

Across the centuries, from the roof models of the master carpenter Hugh Herland in the 1390s to NASA's space station mockups of the 1990s, builders of pioneering works have hedged their bets on the performance of their yet-unbuilt masterworks in every available way. After 600 years the device still favored to provide a final review of the new thing's design is a three-dimensional model that supplies designers and builders with nonverbal, sensual, qualitative information—visual, tactile, muscular, and aural.

Appendix: The Hazards of Copying Technical Drawings
Figures 4.23 and 4.25 show drawings made by Francesco di Giorgio around 1470. These three drawings were copied in the 1540s by Sienese artists who had been trained to copy technical drawings. Despite their training, the copyists introduced significant technical changes as they made their copies.[32]

Figure 4.24 shows the copy of figure 4.23. The horizontal screw at top has lost some of its threads and its attachment to the rolling carriage. Turning the large cylindrical nut on the horizontal screw of Francesco's original would move the rolling carriage to the right or the left as the nut pressed against the left or the right stop; turning the nut on the copy would move the screw but not the carriage. Francesco's winch at the base of the crane's central column has a horizontal rope drum and a crown gear on the drum's axis. The crown gear was to be turned by the square nut attached to the lantern pinion, thus gaining a mechanical advantage. The copy has a vertical rope drum with a square nut on its axis. There is no mechanical advantage as the nut is turned, and the crown gear is useless. The frame holding the rope drum rests on a narrow triangular base rather than an ample rectangular base, as in Francesco's original drawing.

Figure 4.26 shows the copy of figure 4.25. Francesco's carriage was to be steered by moving the axle whose rectangular ends slide in slots on the

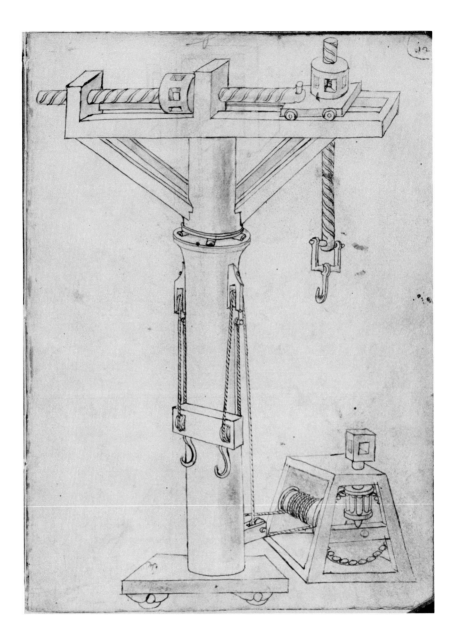

Figure 4.23
Original drawing by Francesco di Giorgio.

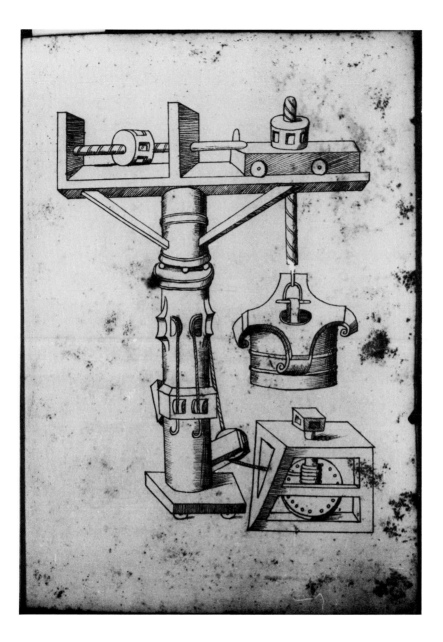

Figure 4.24
Copyist's version of figure 4.23.

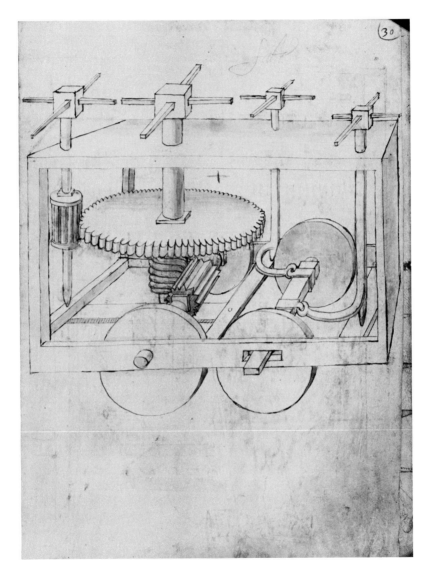

Figure 4.25
Original drawing by Francesco di Giorgio.

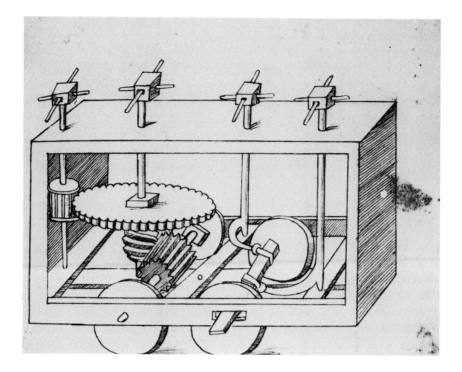

Figure 4.26
Copyist's version of figure 4.25.

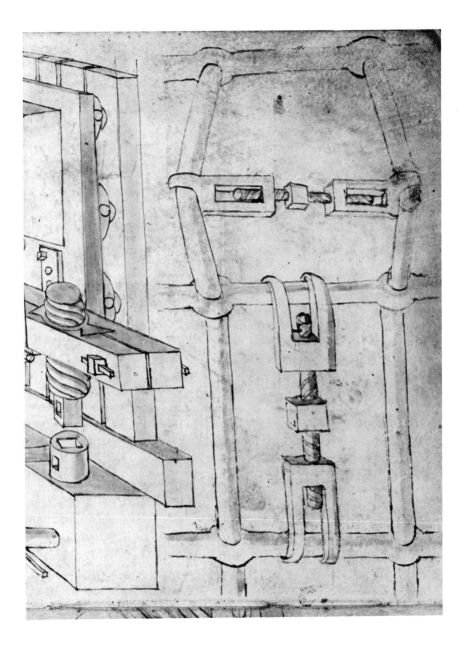

Figure 4.27
Original drawing by Francesco di Giorgio.

near side of the carriage (and presumably on the far side also). The near half of Francesco's movable axle is attached to the J-shaped end of a capstan's vertical turning shaft. Note carefully the attachment at the end of the J, a loop fitting loosely in a loop attached firmly to the axle. The attachment on the far half of the axle is also a pair of loosely fitting loops. On the other hand, the copy shows two solid attachments instead of loosely fitting loops. Francesco's arrangement was clumsy, but it would work; the copyist's version would not.

In his own drawing (Figure 4.27), Francesco's bar spreader and bar bender are similar to modern turnbuckles, which push ends apart or pull ends together when the square nuts at their center are turned. The screws have right-handed threads on one side of the nut and left-handed threads on the other; the threads in the copy shown in figure 4.28 are left-handed throughout the length of each screw. The differences are slight but crucial. Francesco's devices would work; those of the copyist would not.

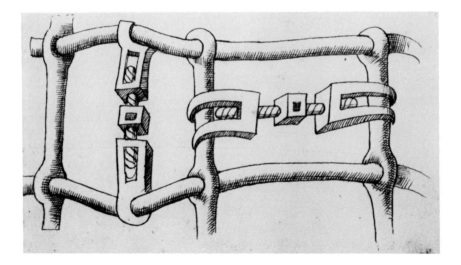

Figure 4.28
Copyist's version of drawing by Francesco di Giorgio.

5 The Development and Dissemination of Engineering Knowledge

Any collection of related facts is difficult to grasp when expressed by figures in tabular form, but the same may be seen at a glance when presented by one of the many graphic representations of those ideas.

Gardner C. Anthony, 1922[1]

In 1871, Thomas Edison was preparing to patent his "automatic printing telegraph," a system that used "neither dots nor dashes" but automatically printed on paper tape the message received. He was busily converting his ideas to drawings on paper. The device that recorded the message, a "translating printing machine," employed a ratchet to advance intermittently the paper tape under a printing head. In order to make his patent claim as inclusive as possible, Edison wrote: "I do not wish to confine myself to any particular translating printing machine, as I have innumerable machines in my mind now which I shall illustrate day by day when I have the spare time."[2] A few of the "innumerable machines" that were in Edison's head are shown in figure 5.1. These sketches say less about Edison's inventive ability than they do about the many alternative solutions to mundane mechanical problems that Edison, like any other avid reader of the technical books and periodicals of his day, had seen and had added to his visual memory. Edison's sketches were not copied directly from Henry T. Brown's 1868 book *Five Hundred and Seven Mechanical Movements,* which records a sample of the mechanisms widely known in the industrial nations. (See figure 5.2.) He was, however, familiar with their principles, as is evident from his sketches.

The fact is that in Edison's day any of a very large number of engineers and other technical people could have promised to devise numerous mechanisms for the intermittent advancement of a paper tape because there were many published drawings of various ratchet devices.[3]

To trace the origins and the development of the vast body of general visual technical knowledge that exists today is one purpose of this chapter. The other purpose is to show how models and other graphic tools of visual knowledge have been employed to teach and use Newtonian mechanics, hydraulics, and allied subjects.

Renaissance Picture Books
The number of technologists whose minds could be engaged by a particular problem or stimulated by a particular idea was greatly enlarged by the appearance of illustrated printed books. Two traditions of such books emerged. The first originated in the engineers' notebooks of the fifteenth and sixteenth centuries and was carried on in printed works such as the heavily illustrated machine books called "theaters of machines." This tradition was simultaneously disruptive and progressive because it suggested new and novel ideas to anyone who could "read" the illustrations. The seeds of the explosive expansion of technology in the West lie in books such as these. The second tradition concentrated on existing technical processes, as in Georg Agricola's classic 1556 study of mining and metallurgical processes. This tradition diffused established techniques but did not promote radical change.

The first "theater of machines" was that of Jacques Besson, published in 1578 in Lyon. Besson's book was followed by that of Jean Errard de Bar-le-Duc, published in 1584 in Nancy. Each contained about 60 engraved plates of mathematical instruments and mechanical devices, such as the sawmill shown here in figure 4.2.[4] The third such volume, by Agostino Ramelli, set the standard for "theaters of machines" for more than a hundred years.

Ramelli was an Italian military engineer in the service of the king of France when his *Le Diverse et Artificiose Machine* [*Diverse and Ingenious Machines*] was published, in 1588.[5] The book is an imposing example of the printer's and the engraver's arts, containing 195 6 × 9-inch copper-plate engravings of grain mills, sawmills, cranes, water-raising machines, and siege devices such as military bridges and hurling engines. A steady stream of such "theaters of machines" coursed through the seventeenth century,

116

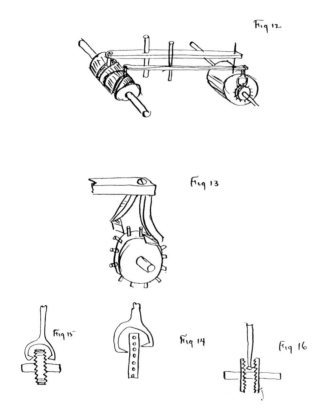

Instead of a worm and gear to drive the paper carrying wheel ahead
The idea has occurred to me that a cam motion the same as
shown in figure 4. could be used thus fig 12

Fig 12

Fig 13

Fig 15

Fig 14

Fig 16

Fig 13 shows the manner of feeding the pin wheel ahead step by step —
Fig 15 shows another device for feeding the paper carrying drum ahead
step by step, twice for, at every revolution of the main driving shaft
Fig 14 shows another view of fig 13

Figure 5.1
Sketches by Thomas Edison for a mechanism to advance paper tape under a printing
head in intermittent fashion.[46]

Chapter 5

Fig 16 shows another device - Fig 17 shows another device for getting motion from the main shaft and also for driving the paper drum

Fig 17

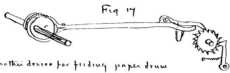

Fig 18 Is another device for feeding paper drum

Fig 18

Fig 19 The same

Fig .19 .

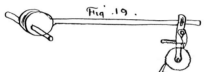

Fig 20 Another device for procuring a ___ double motion for each revolution of the driving shaft for feeding the paper drum =

Fig 20.

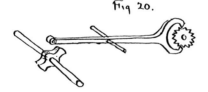

Fig 21 Represents a case for keeping the chemically prepared paper in to prevent evaporation, and to keep it wet and ready for use the moment the machine commences operation, heretofore in Chemical Telegraph. considerable trouble has been experienced in keeping the paper sufficiently damp to give a clear record, and I find that no provisions for exterminating this difficulty have been yet made to my Knowledge. In my machine more than any other it is absolutely necessary that the paper should be in the machine and always ready for use. and owing to the slowness of the paper in passing

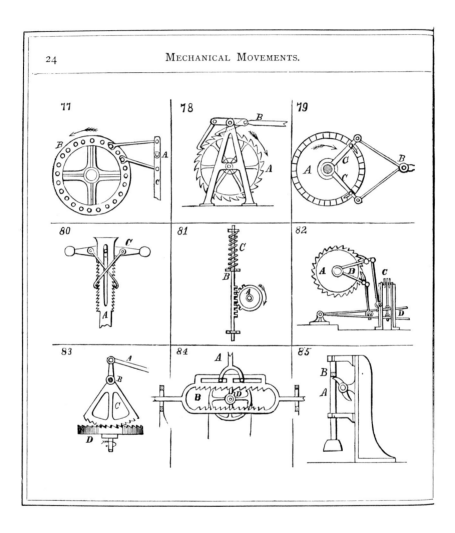

Figure 5.2
Diagrams from the book *Five Hundred and Seven Mechanical Movements*, by Henry T. Brown.[f47]

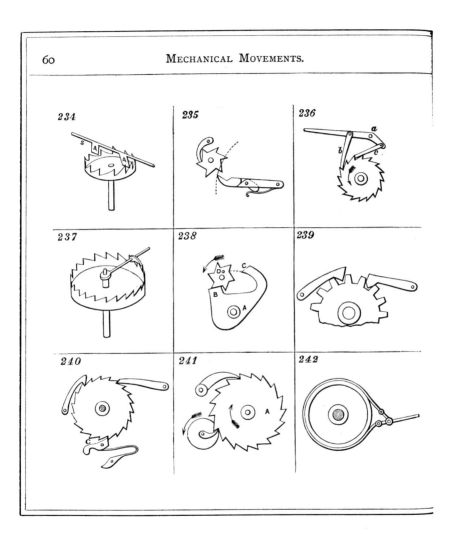

each book adding new ideas and reinforcing ideas already expressed in its predecessors.[6]

Yet by the time of Ramelli's book, in 1588, there was already a body of technological inventions that far exceeded society's demands or needs. As one leafs through Ramelli's hundred varieties of water-raising machines, the conviction grows that Ramelli was answering questions that had never been asked and solving problems that nobody but he (or perhaps another engineer) would have posed. There is no suggestion that economic forces induced those inventions. The machines were clearly ends, not means. Nevertheless, nearly every one of Ramelli's machines, however elaborate or extravagant, has been put to some use in succeeding centuries. (See, for example, figures 5.3–5.5.) Ramelli and his colleagues, supported in their imaginative excursions by their royal and aristocratic patrons, were in fact happily compiling pictorial catalogues of material progress and technical possibilities.

Early in the eighteenth century, the cycle of Renaissance picture books was brought to a close by Jacob Leupold, a German instrument maker, who published a monumental series of machine books entitled *Theatrum Machinarum*. Published between 1724 and 1739, the work collected and distilled in 10 volumes and nearly 500 engraved plates the mechanical repertoire that had accumulated by that time.[7]

Leupold's books and many others of the seventeenth and eighteenth centuries were pored over in the early nineteenth century by Jean N. Hachette and others at the Ecole Polytechnique who were bent upon classifying mechanical devices by function. They made small schematic drawings of each mechanism and displayed several dozen on a sheet (figure 5.6). That scheme of classification was modified by many individuals in Europe and the United States, resulting in numerous charts of "mechanical movements" that gained wide popularity in the nineteenth century (e.g., figure 5.2). In the twentieth century, panels of moving gears, cams, and linkages, powered by small electric motors, were built for use in engineering schools (figure 5.7).

Throughout the many theaters of machines, and in more recent illustrated technical works, a reader finds continuing echoes of Renaissance engineers' notebooks. Recognizable elements of Mariano Taccola's and

Francesco di Giorgio's drawings of the fifteenth century, for example, reverberate in books published in Italy, France, England, and the United States from the fifteenth century onward. Figures 5.8–5.13 illustrate the persistence of a "generic" mechanical idea—a bellows pump, operated by leaning first to one side then the other—from 1451 to 1943. Thus the disruptive and progressive tradition supplied a basic part of the working repertoire of disruptive and progressive inventors of the nineteenth and twentieth centuries.

The second tradition of transmitting technical information through illustrations was established in the middle of the sixteenth century with books by Biringuccio and Agricola on the mining and refining of metals and other minerals.[8] Both authors filled their books with solid technical information and current best practice. Both books contain extensive, detailed descriptions of various processes—assaying, mining, smelting, refining, and founding—and numerous illustrations. Biringuccio's *Pirotechnia* (1540) includes about 85 simple but informative wood engravings; Agricola's *De Re Metallica* (1556) has over 250 wood engravings, many quite elaborate (e.g., figure 4.7 above). Agricola, calling attention to the many illustrations in his book, wrote in the preface: ". . . with regard to veins, tools, vessels, sluices, machines, and furnaces, I have not only described them, but have also hired illustrators to delineate their forms, lest descriptions which are conveyed by words should either not be understood by men of our own times, or should cause difficulty to posterity."

The books that succeeded *Pirotechnia* and *De Re Metallica* were generally modest but solid works useful to artisans engaged in mining, refining, and metalworking. The first books on smelting and assaying were published in Germany before the end of the sixteenth century; others, describing the hardening and tempering of steel, the melting, casting, blanking, and striking of gold and silver coins, and the tensile testing of wire, appeared in the seventeenth century.[9] A notable contribution to manuals for artisans in the mechanical trades was made by the Englishman Joseph Moxon, who in 1683 commenced publishing, in parts, *Mechanick-Exercises, or the Doctrine of Handy-works*, describing the trades of printing, smithing, joinery, turning, and bricklaying.[10]

Figure 5.3
Agostino Ramelli's sliding-vane rotary pump, illustrated in his machine book, was an
original invention. As a waterwheel turned the slotted rotor of the pump, four sliding
vanes moved in and out, maintaining contact with the casing and impelling successive
slugs of water through the discharge pipe.

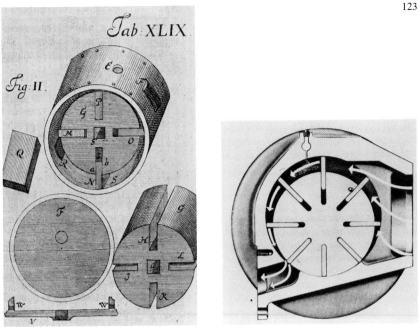

Figure 5.4
Ramelli's pump was copied by Jacob Leupold in 1724. Leupold remarked on the excessive friction in the sliding vanes, pressed on one side by the water being pumped. Problems of friction and a high-speed power source delayed the oil-flooded sliding-vane rotary compressor (figure 5.5) until the 1950s.[48]

Figure 5.5
Oil-flooded sliding-vane rotary compressor.

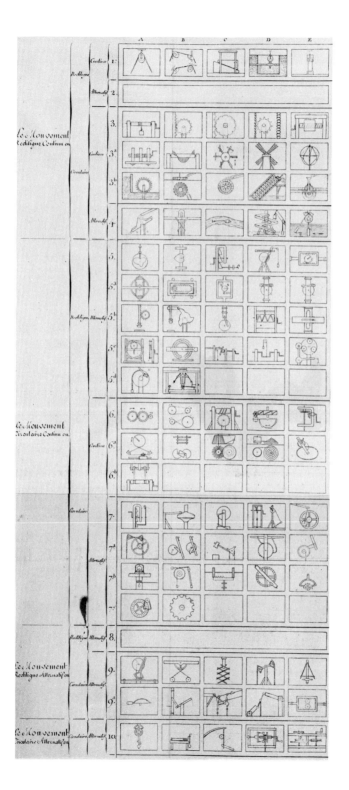

Figure 5.6
Jean Hachette's synoptic chart of elementary mechanisms, 1808. This was the first of many charts of mechanical movements, which enjoyed wide popularity for over 100 years.[49]

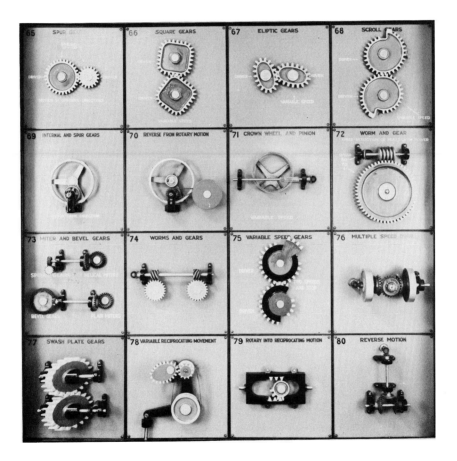

Figure 5.7
A panel of kinematic models by Will M. Clark (1929).

Figure 5.8
Mariano Taccola's drawing of a bellows pump, 1449.[f50]

Figure 5.9
Francesco di Giorgio's drawing of a bellows pump, 1475.[51]

The Development of Engineering Knowledge

Figure 5.10
Vittorio Zonca's engraving of a bellows pump, 1607.[f51]

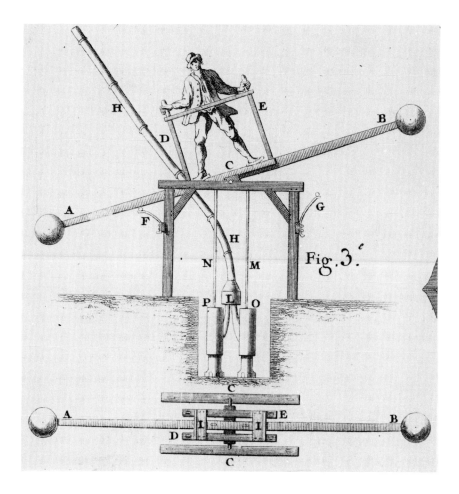

Figure 5.11
Bernard F. de Belidor's water pump, 1739.[f52]

Figure 5.12
Excerpt from a Dictionary of Engineering, 1873.[53]

The tradition of describing actual machines, tools, and processes with words supported by clarifying illustrations culminated in two remarkable French works of the eighteenth century: Diderot's *Encyclopédie* (1751–1780) and the Académie des Sciences' *Descriptions des Arts et Métiers* (1761–1788). Although the books of this second tradition, starting with Biringuccio and Agricola, preserved a remarkable record of extant tools and processes, their influence as agents of progressive change was less than that of the Ramelli-Leupold tradition. An Agricola described what was; a Ramelli described what might be.

The Illiberal Arts

Other intellectual currents in the sixteenth and seventeenth centuries encouraged the visual, nonverbal study of actual machines. A few troublesome thinkers began to insist that anyone who made pronouncements about the built world should have more than a literary knowledge of that world. In 1531, Juan Louis Vives, a philosopher and tutor in the English court, urged scholars to pay attention to the world around them, including the work of artisans, in order that their speculations might be grounded in reality rather than in "foolish dreams." Two years later, in François Rabelais' story of Gargantua, teacher and pupil visited the shops of goldsmiths, watchmakers, and printers to observe artisans at work and thus to complete what Rabelais conceived to be the essentials of a liberal education.[11] Two hundred years later, Benjamin Franklin agreed that as a part of a proper education boys

Figure 5.13
A water pump, 1943.[f54]

should be "suffer'd to spend some time" at "shops of artificers observing their manner of working."[12]

Bernard Pallisy, a noted French potter, believed that so-called learned men should also observe things being made. In his book *Discours admirables* (1580) he invited scholars to come to his workshop, where, he claimed, in only two hours they might see for themselves that "the theories of many philosophers, even the most ancient and famous ones, are erroneous in many points."[13] In 1669, Isaac Newton advised a young friend who was traveling to the Continent to seek out and observe "trades & arts wherein they excell or come short of us in England." In particular, Newton urged his friend to see what the Dutch had achieved in the grinding and polishing of "glasses plane," to inspect military installations, mines, and metal works, and perhaps to visit instrument makers to learn whether pendulum clocks might be used for finding longitude at sea. Newton was especially concerned about his friend's gaining firsthand acquaintance with technological matters.[14]

By Newton's time, the European intellectual community had been deeply influenced by Francis Bacon's grand scheme to enhance "the power and greatness of man" through a new and practical science. As part of a comprehensive plan, Bacon called for a series of "natural histories of trades," intended to study each craft in turn, to describe tools, techniques, and processes, and to make public the technical information that had for so long been known only in workshops. Bacon assumed that the knowledge thus made available would be thought about and improved upon by scholars, that the general level of technical knowledge would be raised, and that progress would be the sure result.

Between 1650 and 1660, enormous amounts of thought and planning were expended by John Evelyn, William Petty, and others on compiling a massive description of trades—in the words of Evelyn, a "History of Arts Illiberal and Mechanical"—for which, Petty explained, "bare words being not sufficient, all instruments and tools must be pictured, and colours added, when the descriptions cannot be made intelligible without them."[15]

Implicit in this project was an assumption that the systematic inductive study of any art by liberal minds would lead to its improvement. It was supposed that new and useful ideas were bound to emerge when ingenious practices in the various trades were studied and compared by men of learning. The new ideas, in Bacon's words, would come about "by a connexion and transferring of the observations of one Arte, to the uses of another, when the experience of several misteries shall fall under consideration of one mans minde."[16] Thomas Sprat, Secretary for the Royal Society, presumed that the histories of trade would encourage the invention of improvements by members of the society: "A large and an unbounded Mind is likely to be the Author of greater Productions, than the calm, obscure, and fetter'd Endeavours of the Mechanics themselves."[17]

Not everyone agreed that the histories of trade would be an unqualified gain for all men. Robert Norman of London, a maker and seller of ships' compasses who called himself an "unlearned mathematician," objected to the overbearing presumption of those "learned in the Sciences" that "Mechanitians," who carry much of their knowledge "at their finger endes," should "be forced to deliver unto them [the learned] their knowledge and conceits [i.e., ingenious plans], that they might flourish upon them, and applye them at their pleasures."[18] Evelyn saw how the histories of trade would "disoblige some, who made their professions their living," and suggested that the detailed knowledge of a man's trade not be made public immediately but instead be explained—"not without an oath of secrecy"—to certain scholars, who would then decide what might be released for the "benefit of the nation."[19]

The compilation of technical information about making things, with which the learned members of the Royal Society had no prior sensual experience, apparently became more laborious than the "benefit of the nation" seemed to warrant. Accordingly, the members soon found other subjects more congenial than the details of workshops, and in the early 1670s the histories of trades were quietly laid aside.[20]

Meanwhile, in Paris, Bacon's program of histories of trades was placed on the agenda of the Académie des Sciences almost as soon as the Académie was formed in 1666. Although Jean Baptiste Colbert, the king's minister concerned with French commercial development, actively promoted a project on "Description des arts et métiers," the first chapter was not written until 1704, and it remained unpublished until mid-century. Work on the

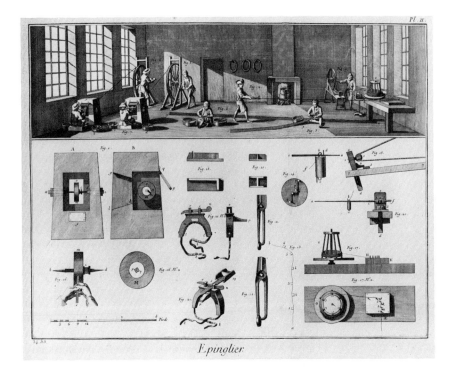

Figure 5.14
Pin-making in France, 1765: workroom and details of tools. Almost certainly, this was the source of Adam Smith's passage in *The Wealth of Nations* on pin-making as an example of the division of labor.[55]

project went forward very slowly; in the 1740s, when Denis Diderot began to make plans for his *Encyclopédie,* he found that some 150 drawings and engravings had been prepared for the "Descriptions" project. Diderot drew heavily on them.[21]

In 1757, after several volumes of the *Encyclopédie* had been published, Duhamel du Monceau revived the Académie's project. Between 1761 and

1788, a handsome folio series of 80 parts was published under the title *Descriptions des Arts et Métiers.*[22]

The French illustrations of workshops and tools are important for the record they provide of eighteenth-century industry, but their significance in stimulating technological change was probably slight. The whole enterprise was considerably removed from the cutting edge of technological innovation. The *Encyclopédie,* in the tradition of Biringuccio and Agricola, supplied a manual for those who wished to read how things were made; however, the roots of radical change are to be found in works (such as charts of mechanical movements) that grew out of the tradition of Ramelli and Leupold.

The Didactic Role of Models

In the seventeenth century, models were used more and more to inform various people of the nature of available machines and devices for carrying out a wide variety of technical enterprises. In 1610, at the Wollaston colliery in Nottinghamshire, there were "models to be seene of all the water-workes [systems for pumping water] that are of any worth or value in Italye, Germanye, or the Low Contryes." As coal mines were driven deeper, the ability to pump out the lower levels with "water-workes" had become crucial to the continued operation of a mine, so an entrepreneur had procured and assembled the models for the information and stimulation of colliers of the region.[23] In 1683, a French inventor held a public display of mechanical models in Rue de la Harpe, centrally located in the Latin Quarter of Paris. According to his catalogue, Jean-Baptiste Picot expected to add four models to his exhibition every two weeks. The models were large—6 feet high—and could be operated. Several were three-dimensional renderings of plates in the machine books of Besson, Ramelli, Böckler, and others. Four were Picot's own inventions, including a gang file-cutting machine and a water-raising machine, which he illustrated in his catalogue and explained in detail.[24]

With the formation of scientific societies in the late seventeenth century, the construction and the display of models were encouraged and to some extent standardized. For example, soon after its establishment in 1666, the

Figure 5.15
Cabinet of models in Académie des Sciences, Paris, 1911. Engraved by Sebastien
Leclerc (1637–1714).[156]

Académie des Sciences in Paris employed modelmakers to develop an elaborate cabinet of models of "various widely used machines" (figure 5.15). The
need for such a model cabinet was attributed to the usefulness—"visible et
palpable"—of the science of mechanics.[25] Members of the Académie and
other gentlemen were encouraged to record their inventions in the annals of
the organization and to augment the basic cabinet by donating models of
their inventions. The models contributed to the cabinet included cranes and
other devices for the movement of heavy objects, floating bridges, clocks,
grain mills, and sawmills, a gang saw for stone, a sail-propelled carriage,
dredges, a quill-pen sharpener, mathematical instruments, a grenade
thrower, and in general a motley and not markedly original assortment of

machines. Eventually seven large volumes, illustrated with more than 500 plates, were published.[26]

No such collection of models was assembled by the Royal Society of London. However, the Royal Society of Arts, organized in 1754, made an effort to promote the practical arts by awarding prizes for original inventions designed to solve particular problems, such as removing dangerous gases from mines, improving agricultural implements, increasing the efficiency of the stocking frame, and making paper of materials other than linen or cotton rags. Contenders for the prizes were required to submit models, and in this way the society gradually acquired an extensive model collection. The "repository" in which it was displayed was occasionally opened to the public.[27]

In Stockholm, the Royal Chamber of Models, established in 1756, ranked high on the itineraries of important visitors to the capital city for many years. In a suite of rooms on the top floor of the old royal palace (see figure 5.16), about 200 models of agricultural, mining, and textile machinery, bridges, dams, and fireplaces were tended and explained to visitors by Captain Carl Knutberg of the Fortification Corps. The collection was aimed at training young people who aspired to mechanical pursuits and at inspiring inventors to make economically useful improvements. Additionally, the models were perceived as providing a record for posterity, so that current accomplishments might be placed in a historical sequence of development. The core of the collection came from Christopher Polhem's "Laboratorium mechanicum," established around 1700 to promote the study of machines that might aid the economic development of Sweden; other models were transferred from the collections of the Board of Mines, the Board of Commerce, the War Office, the Fortification Corps, the Royal Swedish Academy of Sciences, and the Swedish Ironmasters' Association.[28] Of particular interest was a series of models called Polhem's "mechanical alphabet." These were models of mechanical movements described by Carl Cronstedt, a protégé of Polhem, as necessary for a "mechanicus" to know and keep in mind as he designed complex machines (see figures 5.17, 5.18). Polhem saw the five "powers" of Hero of Alexandria—the lever, the wedge, the screw, the pulley, and the winch—as the vowels of his mechanical alphabet. "Not a

Figure 5.16
Surviving models from the eighteenth-century Royal Swedish Chamber of Models as
they appeared in the Tekniska Museet in Stockholm in 1947.

word can be written that does not contain a vowel," he averred; "neither can
any machine limb be put in motion without being dependent on one of
these."[29]

The Conservatoire des Arts et Métiers, a technical museum established
during the French Revolution and located since 1799 in the appropriated
priory of St. Martin of the Fields, depended heavily on models to carry out
its mission as a source of technical information, instruction, and inspiration
for artisans and others. Several exquisite miniature three-dimensional workshops that brought to life the plates of Diderot's *Encyclopédie*, built before
the Revolution for the Duke of Orleans' children, were commandeered by
the museum. Many of the models of Jacques Vaucanson, a builder of autom-

Chapter 5

Figure 5.17
Surviving examples of Christopher Polhem's "mechanical alphabet" in Tekniska
Museet, 1947.

The Development of Engineering Knowledge

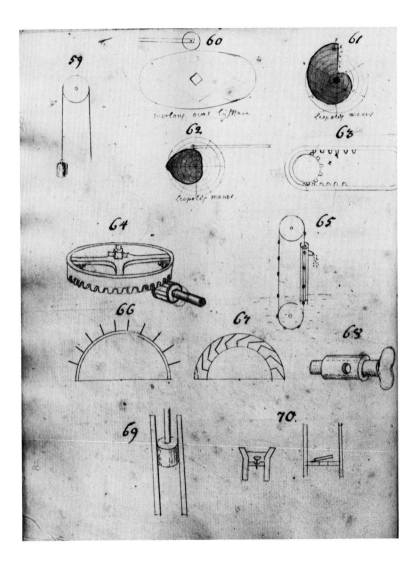

Figure 5.18
Part of Polhem's mechanical alphabet as recorded in Carl Cronstedt's notebook,
c. 1729. Cronstedt noted the sources of diagram 60 (Morland) and diagrams 61 and 62
(Leupold).[57]

Chapter 5

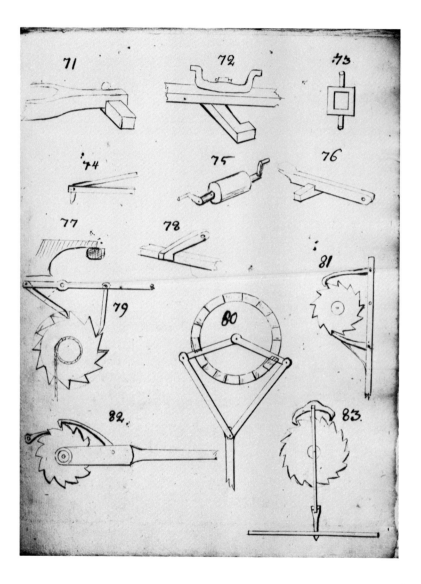

The Development of Engineering Knowledge

ata and machine tools, were added to the Conservatoire's collections after his death.[30]

Probably the largest collection of models ever assembled was that of the United States Patent Office. From 1790 until 1880, every patent application had to be accompanied by a model (no larger than 12 inches square) "if the nature of the invention will admit of a model." In contrast with most of the model collections in Europe, the public had free visual access to the Patent Office's collection. It was undoubtedly visited by interested inventors and others, but it also became, for a time, a popular tourist stop. Surviving pictures show gentlemen, ladies, and children strolling through the galleries in which the models were displayed in glass cases.[31]

Teaching Physical Principles
In early-eighteenth-century Britain many artisans and tradesmen learned the rudiments of Newtonian natural philosophy in the numerous public lecture courses that were given, for a fee, in London and in the growing provincial industrial cities, such as Manchester and Birmingham. These well-attended lecture series explained the principles of mechanics, hydraulics, and aerostatics (basic topics of classroom physics from the eighteenth century until World War II). A considerable number of artisans joined merchants and other members of the new middle class in the audiences, which often numbered several hundred.[32]

Probably the best-known lecturer was John T. Desaguliers (1683–1744), born in France and brought to England at a very early age by his Huguenot parents. He took an M.A. at Oxford, and at 31 he became a Fellow of the Royal Society of London. Desaguliers eventually published his lectures (in two heavily illustrated volumes), as did his fellow lecturers Benjamin Martin and James Ferguson.[33] John Fitch and James Rumsey, early American builders of steamboats, discovered the details of Newcomen's steam engine in those volumes: Fitch in Martin and Ferguson, Rumsey in Desaguliers.[34]

Desaguliers and his colleagues built an elaborate array of apparatus and working models to demonstrate various principles. A few surviving examples that were brought from England to the United States in the 1760s for use in classes at Harvard College exemplify the elegance and the utility of these devices. A set of brass pulleys and weights, mounted in a mahogany frame, was built for Harvard by Benjamin Martin, who also supplied a model of a two-cylinder water pump to show how a fire engine worked (figures 5.19, 5.20). The reciprocating pump with nozzle, similar in principle and arrangement to Hero's pump, was a fitting symbol of the continuity of well-designed contrivances, as well as an example of one workhorse without which the present industrial era is impossible to imagine. A vacuum pump was supplied by another London builder.[35]

Models of elements of machines were used extensively in European and American engineering schools until the Second World War. Franz Reuleaux, German author of the 1875 treatise that established the kinematics of machinery as an engineering subject, designed a set of nearly 300 models of gears, cam, linkages, crank mechanisms, and many other "mechanical movements" as a supplement to his textbook. Reuleaux's models, sturdily constructed of iron and brass, were sold to engineering schools in Europe and the United States.[36] (See figures 5.21 and 5.22.)

The Tools of Visual Analysis

A cluster of powerful graphic ideas that came together early in the seventeenth century formed the visual basis of most mathematical calculations in modern engineering. Napierian logarithms and Cartesian coordinate geometry, followed later by graphic statics and nomography, enabled engineers to visualize their calculations in ways that they could not with numbers alone.

In 1614, John Napier of Scotland worked out the idea of the logarithm, which is in fact an exponent, as a means to carry out multiplication and division by merely adding or subtracting the exponents. Napier's friend Henry Briggs suggested the decimal base 10 and, upon the death of Napier, calculated the logarithms of 30,000 numbers to 14 decimal places. Briggs arranged his logarithms in tabular form, just as current books of logarithms do.[37] In 1624 the Englishman Edmund Gunter produced the first logarithmic scale, incorporating in it the medieval mathematician Nicole Oresme's idea of representing a number by a distance along a straight line. The idea of the slide rule, in which one Gunter scale slides past another, visually adding or

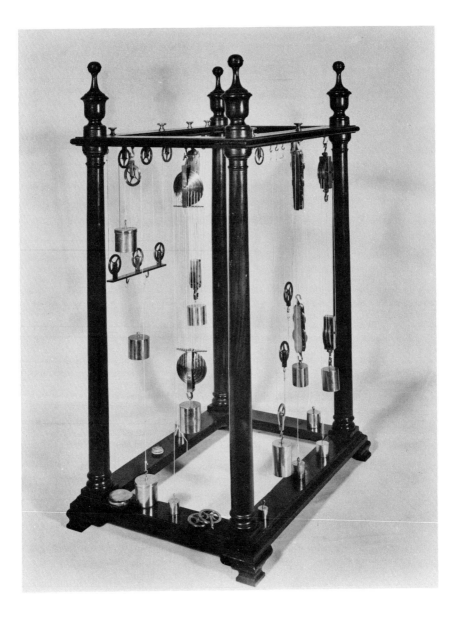

Figure 5.19
Harvard teaching apparatus, 1765–1900 + : mahogany stand with pulleys and weights
demonstrating principles of Newtonian mechanics.

Figure 5.20
A two-cylinder force pump with a nozzle, used to demonstrate the principles of the fire pump.[58]

subtracting two logarithms, belongs to William Oughtred, another En- glishman, who in 1632 also invented a circular slide rule.[38]

At first Oughtred's slide rules were of little utility, because one could read from them only approximate answers. Their utility increased as the developing engineering sciences—mechanics, hydraulics, and so on—en- couraged more and more numerical calculations. Although it is a straightfor- ward operation to copy logarithms out of a table, to add or subtract them, and to refer again to the table for the result, frequent repetition is tedious. Slide rules give answers that are "good enough" for engineering work—that

Figure 5.21
Cornell College's "Museum of Mechanisms" in 1885. A collection of Reuleaux kinematic models is shown at left; in the right foreground are models of governors, steam engines, and other mechanical devices.[59]

is, the results of slide-rule calculations are as precise as the quantities that an engineer works with.

The slide rule was the prime symbol of the engineering profession until the 1960s, after which it was made obsolescent by digital computers. Current computer calculations, yielding a dozen or more significant figures, are more precise than slide-rule calculations, which yield but three significant figures, but they are seldom more accurate. Most of the data used in engineering are, by nature, approximate. In general, the precision and the speed of the computer are bought at the cost of the visual sense of the reasonableness of

$$\frac{n_1}{n} = 1 - \frac{a\,c}{b\,d}.$$

Figure 5.22
One of the Reuleaux models, an epicyclic gear train.[f60]

The Development of Engineering Knowledge

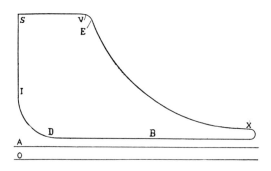

Figure 5.23
An "ideal" indicator card.[61] (See note 39.)

a numerical answer that many engineers cultivated as they learned to make calculations on their slide rules.

A second visual system of calculation also dates from the first half of the seventeenth century. René Descartes developed a system for the graphical representation of quantities that both figuratively and literally added a new dimension to Oresme's numerical scale. Descartes' contribution was to locate a point in a plane by noting its distance from a pair of intersecting lines, or axes. The intersecting lines, carrying scales of distance, provide a coordinate system to locate points and lines. In modern terms, the ordinary "curve" (which may also be a straight line) on Cartesian coordinates is the trace of a mathematical relationship between the properties of a body, a device, or a system. The mathematical relationship thus becomes a visual image of great utility. Characteristic curves that describe the performance of machines or materials provide easily remembered shapes relating changing variables. If the efficiency of an internal-combustion engine is at issue, for example, a mechanical engineer's thoughts turn instinctively to a performance curve (typically one of efficiency versus load); a structural engineer carries a mental library of stress-verus-strain curves; and other engineers, using appropriate curves, visualize performance characteristics of their systems under specified conditions.

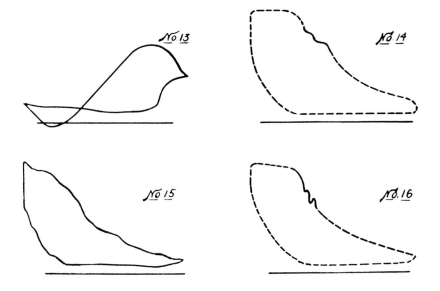

Figure 5.24
Indicator cards showing deviations from the ideal.[62] (See note 39.)

The Cartesian coordinate system may also include a third axis, perpendicular to the other axes already described, that allows the location of points or lines in space by means of three coordinates. Cartesian coordinates utilize visual images to clarify relationships among abstract concepts in engineering, making possible diagrams of, for example, a system of forces or a pressure-volume-temperature "surface" of a perfect gas.

Analytical geometry, also based on Cartesian coordinates, is in effect a visual algebra that accommodates curves of formal equations and also empirical curves described by no simple equation, such as an indicator diagram of a reciprocating engine.[39] (See figures 5.23 and 5.24.) Analytical geometry led to geometric calculus, which represents differentials as instantaneous slopes of curves and integrals as areas under curves.

The visual logarithms of Gunter and Oughtred, the Cartesian coordinate system, and the visual algebra and calculus of Descartes provided a versatile foundation for the entire range of visual systems of engineering calculations that appeared in the nineteenth century.

In the 1860s, Karl Culmann, an instructor at the Federal Technical Institute in Zurich, developed a coherent system of graphic statics that was adopted by many engineers throughout the world. Using Culmann's method, one could calculate the stresses in trusses for roofs and bridges graphically with justifiable confidence. Under the assumption of an equilibrium of forces at every joint in a truss (a basic requirement of any stationary structure), a graphical construction called a *force polygon* made it possible to predict relative forces in the members for a given loading of the structure. Forces might be read directly from the force polygon, being proportional to the lengths of lines in the polygon.[40]

The advantages of graphical statics are qualitative, presenting in the calculations a sense of "what's going on"—a "feel"—and permitting the engineer to build in the mind's eye a vision of the forces in a complex structure. Elwin C. Robison, a colleague of mine who trained as an engineer before becoming an architectural historian, related to me an incident in which he was asked to analyze a roof truss of a building to which an addition was planned. His supervisor suggested that he use graphic statics. Robison's response to the task and the way he finally carried it out can best be under-stood in his own words:

> When given the assignment to analyze the roof truss, I immediately reached for my calculator. As an engineering student of the calculator era, I was trained to do problems only analytically, although graphic analysis was mentioned in passing in my statics class. The technique caught my imagination and I spent an hour going through an example in the text. With only this sketchy background I resisted [the supervisor's] request that I analyze the truss graphically—besides, why would somebody be satisfied with a three-place graphical analysis when my calculator has 13 internal places? At his insistence I grudgingly complied, probably muttering something under my breath about his fear of technology. However, as I referred back to the textbook and reviewed the technique, I realized that it provided a built-in check of the analysis through the closure of the diagram, and that the magnitudes of forces are visible at a glance. Compres-

sion and tension were clearly distinguished as I worked my way up and down the diagram, and it didn't take long to realize that the three-place accuracy of graphic analysis is far more precise than what you're going to get in the field when the trusses are set in place.[41]

Another visual mathematical technique of wide utility emerged near the end of the nineteenth century in the Ecole des Ponts et Chaussées. Maurice d'Ocagne (1862–1938), a graduate of Ecole Polytechnique, brought together a number of specialized graphical methods to solve particular mathematical equations and developed a generalized approach. He called the family of graphs he designed *nomograms* and their theory *nomography*.[42]

One familiar two-variable nomogram is a Fahrenheit/Celsius temperature converter. At 0°C one reads 32°F; 100°F corresponds to 37.7°C; and all

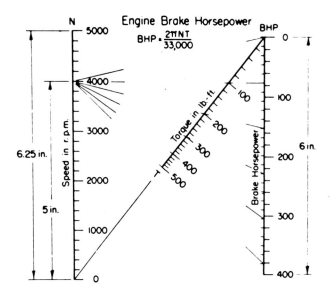

Figure 5.25
A nomogram used to calculate the brake horsepower of an engine being tested with a Prony brake.[63]

intermediate values can be read directly from the scales. The equation upon which this nomogram is based is

$$°F = (\%_5 × °C) + 32.$$

The scales can be extended indefinitely in either direction. The limits of a nomogram are usually decided on practical grounds by its maker.

Nomograms of more than two variables may be constructed with appropriate scales so aligned that a straightedge laid between two variables will indicate specific values of the other variables (figure 5.25). The nomogram serves as a handy repository of all solutions of a given equation between the limits of the selected scales.[43]

When nearly all engineers used slide rules, carried out structural analyses using graphic statics, and resorted to whatever nomograms they might find to solve immediate problems, the advantages of visually monitoring one's calculations (Does it look right? Are the numerical answers reasonable?) were built into the graphical mathematics they used. Even though digital computers are making graphical methods seem both old-fashioned and insufferably slow, a few younger engineers, along with the old fogeys, are beginning to understand that speed has sometimes been bought at the cost of understanding.

6 The Making of an Engineer

These men, who knew no language but algebra, who thought that one was ready for anything when he knew algebra, who esteemed a man only to the extent that he knew algebra, who were incapable of rendering services to this country other than in algebra. . . .

Theodore Olivier on Laplace, Poisson,
and Cauchy, who in 1816 reorganized
the Ecole Polytechnique[1]

The need for some faculty members with actual engineering experience and the desirability of having faculty members who enjoy teaching has not been sufficiently emphasized in recent years.

Frederick H. Kobloss, 1991[2]

The training of Filippo Brunelleschi,[3] Francesco di Giorgio, and Leonardo da Vinci included apprenticeships in which they learned how to prepare and use the materials required to make drawings, paint pictures, and produce sculptures in stone and metal. Their knowledge was based on sensual observations, and they were guided by masters who showed the apprentices what to look for. They were trained as artisans, which today means "persons skilled in an applied art."

When, in the fifteenth century, Brunelleschi designed and directed the building of the great dome of the Florence Cathedral, he demonstrated his knowledge of the properties and behavior of materials, and of the mechanics of an intricate, monumental structure intended to last a long time. It is difficult for a twentieth-century engineer to believe that Brunelleschi could accomplish such an impressive achievement without some help from science. Yet Brunelleschi's knowledge was developed in a world of art, not science.

In the Renaissance books called "theaters of machines," which displayed the latest technological marvels, we might expect to see the symbiotic relationship between art and technology made explicit, or at least acknowledged. On the contrary, a number of enthusiastic engineers of the Renaissance insisted that the mechanic arts had been brought to their current state by the power of mathematics. Ramelli, for example, devoted eight large folio pages in the preface of his machine book of 1588 to convincing his readers that all mechanic arts rest upon mathematics, the "supreme science." Yet his images and his text offer no evidence of his use of geometry, arithmetic, or any other branch of mathematics or formal mechanics in designing his machines. Close study of his drawings reveals his intimate knowledge of mechanical principles as derived from experience in workshop and field. They exhibit a restless originality that is manifested both in totally new devices (for example the rotary pump) and in numerous detailed variations of common elements (such as fastenings and shaft bearings). Ramelli depicted about forty ways to convert rotary to reciprocating motion, but each device was buried in a complete machine. Ramelli, unlike Leonardo, did not recognize such devices as forming a class. Despite his discourse on the "excellence" of mathematics and its universal utility, Ramelli's intellectual resources were not derived from the mathematics of his day.[4]

Lynn White observed the same phenomenon in "every sixteenth-century treatise on the art of fortification," all of whose authors insisted that mathematics was the "essence" of their art. White concluded that "pure calculation was seldom possible" and that "despite all the talk about theory, their methods were essentially empirical."[5]

The authors of books on machines and treatises on fortification apparently recognized the high intellectual status enjoyed by the rediscovered Greek mathematics among the patrons and the humanist scholars who set the tone for acceptable learning during the Renaissance. Then, as now, those "selling" new technical ideas employed a familiar and accepted myth that explained the superiority of the new ideas. Then it was mathematics; now it is science.

The origins of today's science are pertinent to our understanding of the way it is being used today to lend legitimacy to the accepted pattern of engineering education. Starting in 1605, Francis Bacon developed his program for a practical new science that would liberate human society from the stifling old science of bookish assertion and disputation, unleavened by contact with the natural world or the world of work. Bacon envisioned an experimental science resting upon direct observations of nature and incorporating the knowledge and the ingenuity found in the arts and the crafts.[6]

It was in 1620, the year of the Pilgrims' migration to New England, that Bacon called for the writing of a history of trades that would serve as a foundation of his new science. Four years later, he announced in his utopian essay *New Atlantis* his plan for "Salomon's House" (an establishment that we might term a research laboratory), in which "the true nature of things" would be determined through experiments. The details of experiments would be decided upon by various panels of scholars; as the experiments were carried out, a group of "Compilers" would make observations and ascertain principles from the experiments. A group of "Benefactors" would draw from the experiments "things for use and practice for man's life and knowledge" and would deliver to the public the stuff of progress, the technical information that would enhance the power and greatness of humankind. Next, a group of people called "Lamps," after consultation with the Compilers and the Benefactors, would devise new experiments that would permit wider generalizations, "more penetrating into nature than the former." Finally, a panel of "Interpreters of Nature" would "raise the former discoveries by experiments into greater observations, axioms, and aphorisms."[7]

From Bacon's time to the present—more than 350 years—promoters of the mathematical sciences have convinced their patrons that science is *the* way to truth and that it is also the chief source of the progressive inventions that have changed the material world. The myth that the knowledge incorporated in any invention must originate in science is now accepted in Western culture as an article of faith,[8] and the science policies of nations rest on that faith.

We have become so accustomed to this subordination of technology to science that it is difficult to realize that the Renaissance engineer, trained as an artist and retaining the artist's use of intuition and nonverbal thought, had significant counterparts in the United States as recently as the nineteenth century. For example, Robert Fulton of steamboat fame and Samuel Morse, inventor of the American telegraph,[9] were both artists before they turned to careers in technology.

The organization of US technology in the first half of the nineteenth century followed the pattern set in the world of art, although the connections had escaped historians until they were recognized by Brooke Hindle.[10] American societies for the promotion of the mechanical arts were strongly influenced by the London Society of Arts, which encouraged both the fine arts and the mechanical arts through exhibitions and awards. Philadelphia's Franklin Institute, organized in 1824, awarded prizes at annual exhibitions (a practice borrowed from art shows) and sponsored classes in mechanical and architectural drawing. The American Institute of the City of New York, founded in 1828, held annual industrial fairs and, like the London Society of Arts, maintained a permanent exhibition of machines and models for observation and study by mechanics and inventors.[11]

In the latter half of the nineteenth century, the ties between science and technology were strengthened as the principles of thermodynamics provided explanatory theories that eventually led to improvements in the performance and the efficiency of steam engines.

The physical sciences were strongly promoted by the British Association for the Advancement of Science (founded in 1831), the American Association for the Advancement of Science (started in 1848), and the numerous societies of individual disciplines (for example, the American Chemical Society, founded in 1876 by John William Draper, a chemist notable for his influence in popularizing science[12]).

The Progressive era (ca. 1905–1925) was marked by a growing faith in science on the part of many educated citizens who believed that scientists—expert, efficient, and disinterested—could ameliorate "the worst of an industrial society's inequities without calling into question the essential structure of this society"[13] by attacking a series of limited and solvable problems. Throughout the first half of the twentieth century, an increasing number of popular magazines and daily newspapers featured new scientific discoveries and portrayed scientists as selfless public benefactors.[14]

American engineers enjoyed a fair share of this public esteem in the nineteenth century and the early twentieth as they planned and directed the building of a continent-wide network of railroads and highways, of systems that supplied cities with water and removed their sewage, of basic industries that supplied construction materials such as iron and steel, and of the tall buildings that structural steel made practicable. Herbert Hoover, a mining engineer, was Secretary of Commerce in the 1920s; he encouraged engineering organizations to work for the public well-being and promoted such engineering projects as the St. Lawrence Waterway and Boulder Dam.[15] As president of the United States, however, he saw the status of engineers wane as the Great Depression deepened and engineers were widely blamed for the headlong mechanization of industry that had put many thousands of workers out of jobs.

During World War II, Vannevar Bush, an electrical engineer who had been Dean of Engineering at MIT, directed the Office of Scientific Research and Development (OSRD), a powerful agency reporting directly to President Franklin Roosevelt and exerting great influence on the development of new weapons, such as the proximity fuse and guided missiles. Early in the war Bush found that military officers tolerated scientists but that they held a very low opinion of engineers—particularly the "older military men," who tended to keep Bush's engineers at arm's length. "This is the way things were at first in our relations with the military in our war effort," Bush wrote. "So all OSRD personnel promptly became scientists."[16] Twenty-five years later, Bush wrote that "the business of elevating the scientist to a pedestal probably started with this move, and it has certainly persisted and misled many a youth." "Even recently when we sent the first astronaut to the moon," he noted, "the press hailed it as a great scientific achievement. Of course it was nothing of the sort; it was a marvelously skillful engineering job. Now that there is a National Academy of Engineering, perhaps the title of engineer will recover its just recognition."[17]

Despite Bush's pious expression of hope, the prestige of the National Academy of Engineering has always been several notches below that of the National Academy of Science. In fact, Bush's prescriptions and recommendations that led to the establishment of the National Science Foundation (in

1950) helped to ensure that engineers' problems of status would persist, and that the training of engineers would be deeply skewed by the recognition that science received a better press than engineering.

In his influential 1946 book *Endless Horizons,* Bush both set the tone and supplied the prescription for government support of postwar science and engineering. He set the tone of discourse by ignoring engineers and stressing the importance of basic research by scientists: "Basic research leads to new knowledge. It provides scientific capital. It creates the fund from which the practical applications of knowledge must be drawn. New products and new processes do not appear full-grown. They are founded on new principles and new conceptions, which in turn are painstakingly developed by research in the purest realms of science."[18] Although he thought such research should be paid for by the federal government, Bush wrote that "publicly and privately supported colleges and universities and the endowed research institutes must furnish both the new scientific knowledge and the trained research workers. . . . It is chiefly in these institutions that scientists may work in an atmosphere which is relatively free from the adverse pressure of convention, prejudice, and commercial necessity."[19]

As was suggested above, Bush was the architect of the National Science Foundation, a government agency intended to support financially and coordinate the efforts of a network of "Salomon's Houses" located in leading American universities and endowed laboratories. The purposes of the NSF, in Bush's words, was to "support basic research in nonprofit organizations," to develop "scientific talent in American youth by means of scholarships and fellowships," and to support "long-range research on military matters."[20]

The National Science Foundation was not authorized by Congress until 1950, and a separate military research division was never authorized. Nevertheless, ample funds were supplied immediately by the armed services to keep alive the expectations of a continuing war economy in the better engineering schools and the elite universities.

At the end of the war, many opinion leaders were convinced that America's national security depended on scientific superiority.[21] One popular argument for continuing military support of academic science and engineering was that it would permit the US to continue developing armaments without maintaining a large-scale military research establishment.

For 20 years after World War II, the Office of Naval Research and other military agencies supported and—without serious public criticism—guided the direction of a large portion of the research performed in colleges and universities.[22] The impact of this single patron on the physical sciences was shattering: only research that contributed to war-making was valued. The impact of the patronage on engineering schools was even greater: it radically changed the nature of the curriculum and the outlook on the teaching of engineering students.[23]

The Grinter Report

In 1952, one of the periodic internal reviews of engineering education sponsored by the American Society for Engineering Education was undertaken by a 30-member committee of engineering professors and administrators headed by L. E. Grinter, dean of the graduate school at the University of Florida.

A preliminary report was written in about a year. It noted that "during the lifetime of present faculties the art of engineering has come to depend greatly upon the basic science of engineering," and that therefore new faculty members should have the Ph.D. degree. "Appropriate industrial experience is [also] important in a well-balanced faculty," added the report. Such experience might be considered in selecting or promoting faculty members, "but it should not be a requirement for faculty members with special educational background [i.e., the Ph.D. or the Sc.D.] or with demonstrated ability in research or teaching."[24] A dozen years later, it would become painfully obvious that engineering faculties had become strong in research but were generally unfamiliar with engineering practice, particularly design. Nor did the teachers have the necessary industrial experience to introduce students to the many subtle, unstructured problems of designing, building, operating, and maintaining structures and machines.

In addressing the difference between academic research (for which generous funds were being furnished by government agencies) and actual engineering practice, the Grinter Committee sensibly recommended a "bifurcation" of engineering curricula, so that those who wanted to pursue graduate study would not have to learn engineering art and those who expected to be engineers in the field might learn less engineering science and more engineering art.

In October 1953, the preliminary report of the Grinter Committee was sent for critical review to all of the approximately 175 colleges with accredited curricula. (Accreditation was by a committee sponsored by national engineering societies.) Faculty committees of 122 colleges responded. In the words of the next (interim) report, "the extraordinary response of the institutions in reporting their views resulted in an extended and penetrating analysis that was nationwide in scope."[25] Translated, that statement meant that the suggestion of "bifurcation" was seen as a retrograde measure, and that it evoked in the academic entrepreneurs who by then were caught up in the status game of sponsored research "a nearly universal institutional reaction that engineering curricula should not be subdivided into two functional stems."[26] Bifurcation was a dead issue before the interim report was published.

The final report contained two significant recommendations, which were promptly followed by those schools that had received or hoped to receive large research grants. First, "those courses having a high vocational and skill content" should be eliminated, as should "those primarily attempting to convey engineering art and practice."[27] Thus, shop courses—intended to give students a visual and tactile appreciation of materials and basic processes, such as the welding, casting, and machining of metals—were rapidly dispensed with.[28] Engineering drawing lingered a bit, primarily because many drawing instructors held academic rank and were difficult to fire, but the diminished status of courses in drawing and descriptive geometry was clear to all concerned. The "art and practice" courses—which described the individual components of engineering systems such as steam power plants, electrical networks, and chemical process plants and explained how the components were coordinated in practice, thus providing training in the way engineering had been and was being done—survived only until the Committee's second recommendation could be put in place.

That second recommendation called for courses in "six engineering sciences—mechanics of solids, fluid mechanics, thermodynamics, transfer and rate mechanisms (heat, mass, momentum), electrical theory, and nature and properties of materials." Although the report noted that "few existing curricula contain all six engineering sciences," it asserted that "the [scientific] engineer needs background in all six fields listed."[29]

By no means were all engineering curricula changed immediately, but the gospel of change was unambiguous for the research-oriented engineering schools and for the schools that aspired to join the prosperous group.

The final report also called for courses that would provide "an integrated study of engineering analysis, design, and engineering systems for professional background, planned and carried out to stimulate creative and imaginative thinking, and making full use of the basic and engineering sciences."[30] Traditional "art and practice" courses in engineering design were nowhere mentioned, and the "integrated study" often proved to be exercises in analysis. Rather than learn how engineering design had been done in the past and how it "worked," students were encouraged to be creative and imaginative—both of which abilities may result in better designs *only after* one understands the art itself.

The Crisis in Design

In the late 1950s, the declining ability of engineering graduates to carry out design projects was becoming painfully evident to employers and to older teachers and administrators in engineering schools.

In 1961, Paul Chenea, head of the mechanical engineering department at Purdue, observed that while the engineering sciences had developed rapidly, design had "slipped back." Perhaps that was because the "faculty of several institutions have concluded that this may be an area of knowledge that cannot be successfully taught and hence have taken steps to eliminate an experience in design from the students' program."[31] Perhaps a better explanation of the slipping back was that given by B. R. Teare, Jr., dean of engineering at Carnegie-Mellon University: "Design must compete with the engineering sciences for a place in the curriculum and for academic respectability and prestige."[32]

A 1961 report issued by a faculty committee at MIT fully justified the fears of those who saw a crisis in design. Recent graduates were criticized for "unwillingness and inability" to consider a complete design problem and for preferring to tackle only the fully specified problems that could be solved by analytical methods. This tendency was more marked at the M.S. and Ph.D. levels, where young engineers "feel at home in solving problems which have numerical answers—the kind of problems used in school for teaching analytical techniques." Also, these "young engineers tend to consider problems which do not involve mathematics at least at the level of the calculus as beneath their dignity—something to be turned over to a technician who [is] without the benefit of a higher education."[33]

The MIT report was a remarkably honest and perceptive document that resulted from a four-week, full-time study conducted in 1959 by a committee of eleven faculty members. The committee had invited experienced engineering designers representing ten fields of design (aircraft; machine tools; bridges, tunnels, and airports; diesel engines and gas turbines; electrical machinery; operations research; electronics systems; nuclear submarines; chemical plants; and communications systems) to discuss "what they did, what their problems were, and what methods they used in the solution of these problems."[34] The committee also quite accurately identified aspects of undergraduate engineering instruction that supported or discouraged the attitudes and abilities required of successful engineering designers. For this reason it is important to look at its approach and its conclusions in depth.

The report made a serious attempt to articulate the nature of design, beginning with the recognition that there is no unique or "correct" solution to a problem of design. This recognizes that design is radically different from "the kind of problem used in school in teaching analytical techniques." In order to emphasize the difference, let us look for a moment at the committee's description of the way analytical tools are introduced in engineering schools.

Courses in physics, chemistry, mathematics, and the engineering sciences are all mathematical in form. Students manipulate given data to arrive at numerical answers. In all such courses, "questions asked of students are overwhelmingly what will be called single-answer problems. Reasons for the popularity of this type of problem are not hard to find. Single-answer problems provide an essentially objective standard against which the performance of the student can be measured. . . . The student can be taught a series of logical steps to arrive at the answer. The teacher can measure his own effectiveness by noting the percentage of his students who arrive at the right answer when tested. Because of the existence of an objective standard, people inexperienced in engineering, graduate students for example, can be, and often are, entrusted with teaching."[35] The educational problems posed by these ubiquitous single-answer problems—providing not an objective but an arbitrary standard—are several. There is no need to deal with incomplete or contradictory data; any ambiguity merely represents a mistake in stating the problem. One does not need engineering judgment to solve such problems. "Skepticism and the questioning attitude are not encouraged in this situation," said the report. "Neither the data, the applicability of the method, nor the result are open to question."[36]

Turning to a description of the design process, the committee averred that design "is done essentially in the mind," that drawings are pictorial extensions of the mind (an "external (and reliable) memory"), and that "it is not to be expected that all students are equally endowed with the ability to think pictorially any more than to think mathematically. Somehow educators tend to look upon mathematical ability as a more desirable quality than the ability to think in terms of spatial relations."[37]

Because design is "a conceptual process, one in which at least a fragment of a mental plan is necessary before the process can proceed," the MIT committee noted, "synthesis must come first." Because the object being designed is almost always "far too complex to permit analysis directly," a simplified model must be devised to permit calculation. And although "the use of automatic computation" makes it "possible to use more complicated models," "it should not be assumed without proof that a more complicated model represents the physical device."[38]

The necessary but secondary role of analysis in engineering design was clearly articulated in a passage that captures the way designers use accumulated knowledge:

The designing engineer who remains on the frontiers of engineering finds himself making only a small fraction of his decisions on the basis of numerical analysis. When the problem becomes older and more decisions are based on numbers, he moves on to a new and more difficult field where he again finds that a small fraction of his decisions are based on the kind of analysis taught in engineering schools. This is not said to try to belittle the importance of analysis. Everyone recognizes it as an essential tool of the trained engineer. It does not, however, answer all or even a majority of the questions an engineer must answer in a typical design problem, particularly a new one. It seems unlikely that numerical analysis will ever answer more than a small proportion of these questions. The remainder of the questions must be decided on the basis of *ad hoc* experiment, experience (the art of applying knowledge gained by former experiments on the same or similar problems), logical reasoning and personal preference. The subconscious reasoning process, based on experience, which we call intuition, can play a large part.[39]

When the report addressed the suitability of engineering faculties for the teaching of engineering design and the attitudes that make for successful designers, two telling comments opened the discussion: that "embryo engineers should be taught by engineers"[40] and that "the policy of recruiting our faculty primarily from newly made Ph.D.'s and Sc.D.'s leads to teaching analytical techniques to embryo analytical technicians."[41]

It is odd that the irony of the situation of engineering design at MIT went unacknowledged in this otherwise open discussion. MIT had been in the vanguard of the elite engineering schools and universities that had embraced the patterns of faculty research in esoteric, war-driven fields that had only distant and usually questionable relationships to the needs of a viable, peaceful civilization. The irony of being part of the problem rather than part of the solution apparently was either overlooked or suppressed in the committee's report.

Twenty years later, in 1980, two teachers of design in the engineering department of Cambridge University produced *A Survey of Engineering Design Education in North America, Europe and Japan*. John L. Reddeway and Rachel A. Britton visited nearly fifty engineering schools in France, West Germany, the Netherlands, Norway, the United States, and Canada.[42]

In the United States they found that design and any supporting courses that remained, such as engineering drawing, were held in low esteem. In the field of design they saw "no route to the academic top." Creativity courses had flourished in the late 1960s and the early 1970s, they said, "in the Freshman year, partly to attract students, by making engineering fun, and partly to make use of highly qualified staff who had few students in their specialized fourth year courses." They noted that the Engineers Council for Professional Development, the accrediting body for engineering curricula, demanded that "all engineering courses should contain a considerable design element." Some universities "revised their courses, whilst others have merely revised their opinion on how much design content there already is in each course as given at present."[43]

The system of engineering education that Reddeway and Britton "envied most" was that found in Germany and the Netherlands: "There they have time to teach some of the practice of engineering as well as the fundamentals of engineering science. There are no status problems and there is a strong contact between industry and technical universities. . . . Many professors are appointed after spending over ten years in industry."[44]

In Japan, Reddeway and Britton reported, "it is common to find compulsory machine shop courses (varying from 30 to 60 hours in total) for all mechanical engineers, not just those specializing in Manufacturing Technology. Indeed, at Keio, *all* engineering students, including civil and electrical engineers, were given experience with small student lathes and with the stripping and re-assembly of a small motor-cycle engine in the general education course in the second year."[45] This could provide only a marginal acquaintance with a machine shop and its products, but that is better than no acquaintance.

In 1980 the techniques of computer graphics had developed much more slowly than the first flush of excitement had suggested they would. Furthermore, the equipment required for teaching was expensive; so in 1980 there was no clear indication that CAD-CAM would enter all engineering classrooms in the near future. Nevertheless, Reddeway and Britton sensed the alarming possibility that computer-aided design would reach many engineering students as computer-*automated* design, the computer (actually a workstation) coming with ready-made software that would promise to solve whole classes of problems with a minimum of knowledge on the part of the operator. The student would only have to supply appropriate input data—initial and final states, limiting parameters, and the like—and the program would automatically crank out answers.[46] Even at a most basic level, this would mean that the student might well lack an understanding of the meaning or the implication of the answers. At a more sophisticated level, it would mean that all fundamental decisions about the nature of the design would have been made by the programmer of the ready-made software, and that the programmer's assumptions might have very little relevance to the actual design requirements.

The shape of one possible future for engineering design is suggested by a series of advertisements that appeared in magazines such as *Mechanical Engineering* in 1988 and 1989. They open with a picture showing the Tacoma Narrows Bridge as it was collapsing in 1940, a well-known failure that occurred before the careers of most readers had begun. The caption, in large type, reads "What Went Wrong?" The eye is led to the answer: "They Didn't Have IMAGES-3D," a product of Celestial Software. The visual implication is made explicit in the text: "IMAGES-3D gives engineers the ability to design and analyze safer, stronger structures with unprecedented ease-of-use, accuracy, and reliability." For those who may become confused by the program, which is "menu-driven and fully interactive with automatic prompts," the firm offers a way out: the "Ph.D. Support Hotline." The user of the program can talk to a troubleshooting programmer. Here, to quote the 1959 MIT design study, the support hotline, manned by Ph.D. and Sc.D. faculty types, will be "teaching analytical techniques to embryo analytical technicians."

The B.S. in Engineering Technology
In the mid 1960s, while design was crumbling in many prominent engineering schools under the onslaught of high-status military research, a few less-prosperous engineering schools were beginning to award degrees in "engineering technology." In comparison with the engineering curricula of the research-driven schools, the curricula in engineering technology seemed

to be thirty years out of date, including as they did engineering drawing, laboratories for testing of machines, and "art and practice" courses.[47]

The perception among faculty members at the "technology" schools[48] that there was a need for graduates versed in engineering art as well as science was accurate, and the number of B.S. degrees in engineering technology rose from zero in 1965 to about 12,500 by the mid 1980s. During the same period, the annual number of new B.S. degrees in engineering more than doubled, from 37,000 in 1965 to 78,000 in 1986. The numbers of degrees awarded in both categories have decreased slightly since 1986.[49] The graduates in engineering technology were and are finding industrial jobs, and their schools are firmly established. A new journal, *Engineering Technology,* has been published since 1984, reflecting faculty opinion that the *Journal of Engineering Education*—since 1910 the organ of the American Society of Engineering Education—was not providing adequate coverage of the new technology schools.

Many in the engineering profession have been shaken by the rise of engineering technology schools. It has been particularly irksome to learn that B.S. engineering technologists have been offered jobs comparable to those ordinarily held by B.S. engineers.

Commonwealth Edison Company, the Chicago electric utility, hired its first B.S. technology graduates in 1969 and by 1977 had 57 at work. Stanley W. Anderson, a company spokesman, observed that the engineering technologists were at home in equipment-related jobs, and that their laboratory training was "very useful to [them] in such assignments." A large electrical generating station is a showcase of sophisticated large-scale equipment, and to start such a station for the first time is a daunting challenge to the most seasoned engineer. Anderson summed up his view of the young technologists' strengths and weaknesses as follows:

> The education of the typical technologist appears excellent for construction, starting and operating engineering. He frequently is better than the engineer in solving a practical technical problem. He can early in his career fix a valve or repair a motor. He knows how a fuse cutout or a lightning arrester works. He is not as good as an engineer if a problem is theoretical or analytical. He does

not appear as good in some research activities, or in design or in planning. Thus, an operating company staffed by both technologists and engineers, both of which are appropriately assigned, is better staffed than one staffed by engineers alone.[50]

The licensing boards in some states have erected barriers to prevent engineering technologists (ETs) from being licensed as professional engineers, and the Accrediting Board for Engineering and Technology, successor to the Engineers Council for Professional Development, has warned the ET schools that they may not claim or imply that their graduates are qualified as engineers.[51] Nevertheless, as a leader of the ET movement pointed out in 1987,

> There are students who seriously wish to study engineering technology, institutions who wish to teach it, faculty who wish to develop it, and employers who wish to hire the graduates.
>
> Yet over the years proposals arise within engineering societies and educational institutions to abolish, merge, submerge, rename, or otherwise "help" engineering technology. We in engineering technology are like fish in fear of being rescued from our stream by curious monkeys who imagine us to be cold, damp, and unable to breathe. Such depredations share two common misconceptions. First, that engineering technology is or has a problem, and second, that it is in any group's power to abolish the discipline and start all over.[52]

Perhaps one of the best reasons for the rise of curricula in engineering technology is the following, buried in a paragraph of the 1961 MIT report on engineering design:

> Men trained at second-rate schools were often found to be more willing to attempt the solution of a whole problem than those trained at first-rate schools. In spite of inferior technical education these men often worked into positions of leadership whence they directed the work of those from the "better" schools.[53]

There is no doubt that analysis is easier to teach than design and much tidier than laboratory courses. The real "problem" of engineering education is the implicit acceptance of the notion that high-status analytical courses are superior to those that encourage the student to develop an intuitive "feel" for the incalculable complexity of engineering practice in the real world.[54]

7 The Gap between Promise and Performance

> . . . the ship should not have been asked to demonstrate achievements when she had not had a chance to exhibit her weaknesses.

Robin Higham, 1961, on the fatal crash of the British dirigible R-101 in 1930[1]

> "That is by far the most stable pointing we've seen so far."
> "Except that the target isn't in the aperture."

conversation between Spacelab Mission Control and astronaut, 1990[2]

> Absolute faith in any piece of equipment is absolute folly; expect disaster and have a back-up plan to minimize it; be careful.

Jan Adkins, 1980[3]

The Power of Seeing

Until the 1960s, a student in an American engineering school was expected by his teachers to use his mind's eye to examine things that engineers had designed—to look at them, listen to them, walk around them, and thus to develop an intuitive "feel" for the way the material world works (and sometimes doesn't work). Students developed a sense of form and proportion by drawing and redrawing. They acquired a knowledge of materials in testing laboratories, foundries, and metalworking shops. Students at schools close to industrial cities took field trips to power plants, steel mills, heavy machine shops, automobile assembly plants, and chemical works, where company engineers with operating experience often conveyed insights and helped the students recognize the subtleties of the real world of engineering. Students in schools distant from heavy industry were taken by two or three of their teachers on week-long tours of industrial cities.

The students readily accepted the plant visits as a component of their engineering education, considered by their engineering departments important enough to merit making the arrangements and assigning instructors to go along. The plants being visited, for their part, usually assigned operating engineers, and occasionally senior engineers, to serve as guides.

These young engineers' picture of the material world continued to be enlarged after graduation. As working engineers, they routinely looked carefully at many features of the built world as they expanded and refined their repertoire of nonverbal and tacit knowledge. They also seized opportunities to see unusual structures or machines being erected, and they studied accidents and equipment failures on the spot. For example, in 1949, when prestressed concrete bridges were being introduced from Europe into the United States, 500 civil engineers gathered on a specially built grandstand in Philadelphia, in a drizzling rain, to witness a test to destruction of a 155-foot long prestressed concrete bridge girder, the prototype for the 13 girders to be used in the main span of the Walnut Lane roadway bridge. The test girder, with a depth of 6 feet 7 inches and a top flange width of 4 feet 3 inches, was loaded gradually so that its deflection could be observed and measured. The engineers in charge had estimated the probable breaking strength of the girder and had procured enough weights to break it. On the day of the test, however, the girder surprised and pleased its sponsors as it supported all the weights they had on hand. Two days later, enough steel ingots were assembled to break it. Its deflection in the center when it finally broke was 25 inches, resulting from a stress of more than 10 times its working load.[4] The engineers who attended this test were impressed by it and thenceforth carried with them a visual image of the response of prestressed girders to stress. The Walnut Lane tests contributed to the enthusiastic acceptance of prestressed concrete structures in the United States. In 1950, structural steel was used in one-third of the new highway bridges built in California and reinforced concrete in two-thirds. By 1975, structural steel was used in only 2 percent of new bridges, reinforced concrete in 20 percent, and prestressed concrete in 78 percent.[5]

By the 1980s, engineering curricula had shifted to analytical approaches, so visual and other sensual knowledge of the world seemed much less relevant. Computer programs spewed out wonderfully rapid and precise solutions of obviously complicated problems, making it easy for students and teachers to believe that civilization had at last reached a state in which all technical problems were readily solvable.

As faculties dropped engineering drawing and shop practice from their curricula and deemed plant visits unnecessary, students had no reason to believe that curiosity about the physical meaning of the subjects they were studying was necessary. With the National Science Foundation and the Department of Defense offering apparently unlimited funds for scientific research projects, working knowledge of the material world disappeared from faculty agendas and therefore from student agendas, and the nonverbal, tacit, and intuitive understanding essential to engineering design atrophied. In this new era, with engineering guided by science, the process of design would be freed from messy nonscientific decisions, subtle judgments, and, of course, human error.

Problems of Design
Despite the enormous amounts of effort and treasure that have been poured into creating analytical tools to add rigor and precision to the design of complex systems, a paradox remains. There has been a harrowing succession of flawed designs with fatal results—many of them afflicting the projects of the patron that so clearly saw science as the panacea: the *Challenger,* the *Stark,* the Aegis system in the *Vincennes,* and so on. Those failures exude a strong scent of inexperience or hubris or both and display an apparent ignorance of, or disregard for, the limits of stress in materials and people under chaotic conditions. Successful design still requires the stores of expert tacit knowledge and intuitive "feel" of experience; it requires engineers steeped in an understanding of existing engineering systems as well as in the new systems being designed.

Glitch has entered engineering argot to denote a lapse—a design error—in a computerized system. Glitches, however, are not just infrequent lapses; they occur often enough to raise serious questions about the judgment of the designers of those systems. The general collapse of the AT&T long-distance telephone network on January 15, 1990,[6] was but one example of traumatic failures of all kinds of systems that depend upon programmed decision making and are insulated from the common sense that an individual can bring to bear on an incipient failure.[7]

Able people have thought about the limits of objectivity and rationality—that is, the extent to which engineering design can be made "scientific." Their observations hold some clues to the failures and surprises that have followed from a dogmatic faith in computer models as adequate substitutes for reality. James Gleick, in relating the development of the "new science" of "chaos," points out that computer simulations "break reality into chunks, as many as possible but always too few," and that "a computer model is just a set of arbitrary rules, chosen by programmers."[8] You, the programmer, have the choice, he says: "You can make your model more complex and more faithful to reality, or you can make it simpler and easier to handle."[9] For engineers, a central discovery in the formal study of chaos is that a tiny change in the initial conditions of a dynamic system can result in a major unexpected departure from the calculated final conditions.[10] It was long believed that a highly complex system, such as all automobile traffic in the United States, is in principle fully predictable and thus controllable. "Chaos" has proved this belief wrong. The idea that roads will be safe only when all cars are guided automatically by a control system is a typical but dangerous conceit of engineers who believe that full control of the physical world is possible.

Alan Colquhoun, a British architect, argues convincingly that no matter how rigorously the laws of science are applied to the solution of a design problem, the designer must still have a mental picture of the desired outcome. "[Scientific] laws are not found in nature," he declares. "They are constructs of the human mind; they are models which are valid as long as events do not prove them wrong." Most engineering designs, Colquhoun reminds us, must meet requirements that are logically inconsistent. For example, he writes:

> All the problems of aircraft configuration could not be solved unless there was give-and-take in the application of physical laws. The position of the power unit is a variable, as is the configuration of the wings and tailplane. The position

of one may affect the shape of the other. The application of general laws is a necessary ingredient of form. But it is not a sufficient one for determining the actual configuration.[11]

A successful new design combines formal knowledge and experience and always contains more judgment than certainty. Judgment is brought to bear as the designer responds to the design-in-progress by repeatedly modifying means to reach desired ends. Design is thus a contingent process, subject to changes brought about by conditions that come to the surface after the big decisions have been made. It is also a creative process, in which the designer's imagination is required whenever a contingency occurs—and, as Robert W. Mann, a leader in engineering design education, observes, that a creative process "is, virtually by definition, unpredictable":

> The sequence of steps is never known at the beginning. If it were, the whole process could be accomplished by the computer since the information prerequisite to the computer program would be available. Indeed, the creative process is the process of learning how to accomplish the desired result.[12]

Failures and Other Surprises

Engineering design is usually carried on in an atmosphere of optimistic enthusiasm, tempered by the recognition that every mistake or misjudgment must be rooted out before plans are turned over to the shops for fabrication.

Despite all the care exercised by individuals and all the systems that have been used to ensure that all the choices made in selecting parts and arranging them to work together will be correct, the evidence of faulty judgment shows up again and again in some of the most expensive and (at least on paper or on a computer screen) most carefully designed and tested machines of the twentieth century.

Of course there is nothing new about wrong choices and faulty judgments in engineering design. More than a hundred years ago, George Frost, editor of *Engineering News,* tried to track down the reasons for failures of bridges and buildings in order that civil engineers might learn from the mistakes of others. "We could easily," he wrote, "if we had the facilities, publish the most interesting, the most instructive and the most valued engineering journal in the world, by devoting it to only one particular class of facts, the records of failures. . . . For the whole science of engineering, properly so-called, has been built up from such records."[13]

A *Journal of Failures* such as that envisioned by Frost was never published; however, *Engineering News* and its successors have presented many valuable reports of engineering failures.[14] One of those reports—careful, comprehensive, knowledgeable, and fair to all parties—was published in *Engineering News* just a week after a cantilever railway bridge being built over the St. Lawrence River at Quebec City collapsed on August 29, 1907, killing 74 workmen.[15]

The writer of the report spent several days on and about the wreckage, piecing together the evidence before his eyes and the meager testimony of those who lived through the collapse. His conclusion, later confirmed by the Canadian government's year-long official inquiry, was that one of the compression members in the bottom chord of the landward ("anchor") arm of the cantilever truss buckled and immediately brought the rest of the span down on top of it. As figures 7.3 and 7.4 show, the bridge collapsed without significant movement to one side or the other. The failed member, 57 feet

Figure 7.1
Drawing of steel cantilever railroad bridge being built across St. Lawrence River near Quebec City in 1907.[64]

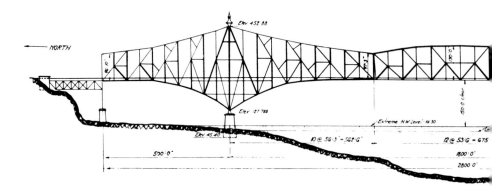

Figure 7.2
The southern end of the Quebec City bridge a few days before its collapse.[f65]

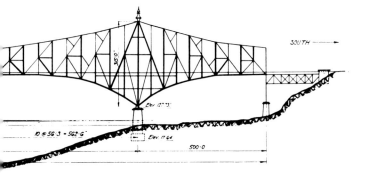

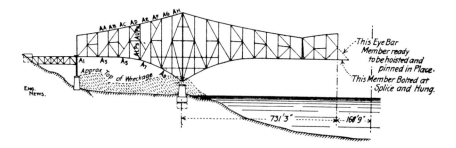

Figure 7.3
The Quebec City bridge fell straight down onto itself, leaving an amazingly compact line of wreckage (as indicated by "Approx. top of wreckage" in this drawing.[66]

long and 4½ feet deep, was of built-up construction, but in hindsight the bracing was not sufficient to prevent the whole member from buckling under lengthwise stress.

"Long and careful inspection of the wreckage," wrote the reporter, "shows that the material was of excellent quality; that the workmanship was remarkably good." But because the members were much larger than those used in ordinary bridges, he questioned the judgment that led to the design of the built-up compression members: "We step up from the ordinary columns of ordinary construction, tried out in multiplied practice, to enormous, heavy, thick-plated pillars of steel, and we apply the same rules. Have we the confirmation of experiment as a warranty? Except in the light of theory, these structures are virtually unknown. We know the material that goes into their make-up, but we do not know the composite, the structure."[16]

Within a few weeks after the Quebec City bridge collapse, *Scientific American* published a "doctored" figure that gave visual meaning to the stress on the bridge's failed chords by setting up one of the chords as a column and perching the USS *Brooklyn,* a 9215-ton cruiser, precariously on it (figure 7.5). In a series of small drawings at the bottom of this figure, the cross section of the Quebec chord was compared visually to the much heavier

Figure 7.4
The low profile of the wreckage of the Quebec City bridge is evident in this photograph. The pile of debris on the bridge pier was perhaps 10 feet above the pier's top surface.[67]

compression chord members of the Hell Gate Bridge (in New York) and a cantilever bridge over the Firth of Forth (in Scotland).[17]

The *Engineering News* report on the Quebec bridge collapse was entitled "The Greatest Engineering Disaster," and in one sense that characterization still holds true. No other construction accident in the twentieth century has claimed as many lives. The runner-up is the 1978 collapse of a "jack-up formwork system" on the top layer of a 168-foot built-up concrete cooling

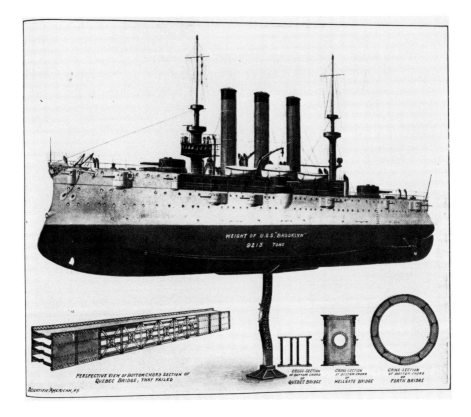

Figure 7.5
Scientific American's lesson on buckling. The weight of the cruiser *Brooklyn* was about the same as the calculated axial load on the chord girder that failed at Quebec City. The perspective drawing of the girder reveals inadequate bracing.[f68]

tower in Saint Mary's, West Virginia. All 51 of the workmen on the tower fell to their deaths.[18]

There is today an unfortunate tendency on the part of many analysts to assume that a failure is the result of incomplete analysis, and that the newer techniques available to designers will prevent similar failures in the future. The report in *Science* of the collapse of a 27-year-old radio telescope, 300 feet in diameter, in Green Bank, West Virginia, implied that such a failure could not occur in a radio telescope designed today. The "cause" of the collapse was pinpointed in "a single highly stressed steel plate" (which had survived for more than 25 years). An "independent panel appointed by the National Science Foundation" declared that "parts of the telescope were under far higher stresses than would be permitted today," and that a "computerized stress analysis would identify potential failure points in telescopes built today, but these methods were not available when the instrument was built in 1962."[19] One wonders what smug and superior-by-hindsight explanation will be given for the collapse, some years hence, of a structure designed today with the help of a "computerized stress analysis."

A much more sensible and realistic outlook on design failures may be found in a little book entitled *To Engineer Is Human: The Role of Failure in Successful Design,* written by Henry Petroski, a professor of civil engineering who graduated from an engineering school in the early 1960s.[20] Toward the end of his book, Petroski has a chapter called "From Slide Rule to Computer: Forgetting How It Used to Be Done." Petroski describes the Keuffel & Esser Log Log Duplex Decitrig slide rule that he purchased when he entered engineering school in 1959 in order to emphasize that the limits of a slide rule's accuracy—generally three significant figures—are no disadvantage because the data on which the calculations depend are seldom better than approximations. Petroski is one of too few academic engineers—he teaches at Duke—who fully appreciate the ambiguities in design and analysis.

Petroski uses the 1978 collapse of the modern "space-frame" roof of the Hartford Coliseum under a moderate snow load as an example of the limitations of computerized design. The roof failed a few hours after a basketball game attended by several thousand people, and providentially

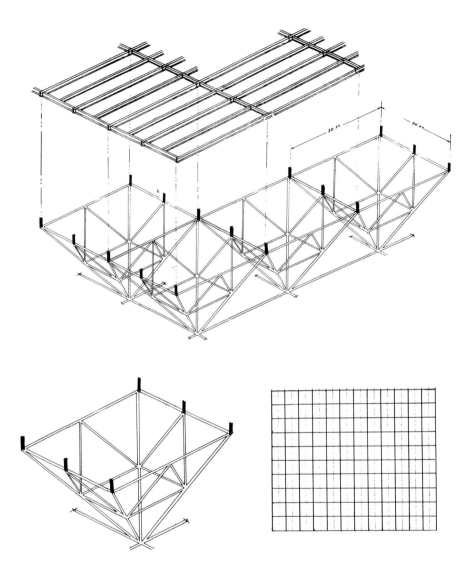

Figure 7.6
The framing system of the Hartford Coliseum. Top: Roof purlin framing. Middle:
Space truss. Botton left: Pyramid module of space truss. Bottom right: Plan view of
pyramid modules.[69]

Figure 7.7
The Hartford Coliseum after its roof collapsed. The contorted corners of the space
frame are visible above the roof level.[170]

nobody was hurt in the collapse. Petroski explains the complexity of a space
frame, which suggests a mammoth Tinker Toy, with long straight steel rods
arranged vertically, horizontally, and diagonally. (Figure 7.6 illustrates the
roof's framing system of space truss and purlins.) To design a space frame
using a slide rule or a mechanical calculator was a laborious process with
too many uncertainties for nearly any engineer, so space frames were seldom
built before computer programs were available. With a computer model,
however, analyses can be made quickly. The computer's apparent precision,
says Petroski—six or more significant figures—can give engineers "an un-
warranted confidence in the validity of the resulting numbers."[21]

In 1979, when a space frame to support the roof of the Gerald R. Ford Museum in Grand Rapids, Michigan, was under construction, Ford became concerned about its safety in view of the collapse in Hartford and one in Kansas City.[22] Accordingly, the roof of the museum was tested by loading it with 200 plastic-lined wooden boxes, each about 12 feet square and filled with water to a level of 8 inches, slightly exceeding the design load of 40 pounds per square foot. The maximum deflection was about 1½ inches, which apparently satisfied Ford and his engineers. The cost of the test, about $40,000, provoked the project architect to comment that "it's unusual to load test a structure of this size because it is obviously quite an expense." The cost of the building was reported as $3 million; the "quite an expense" amounted to less than 1.5 percent of the total cost, a small amount in view of the justifiable concerns about the safety of the structure.[23]

Who makes the computer model of a proposed structure is of more than passing interest. If the model is incorporated in a commercially available analytical program, a designer using such a program will have no easy way of discovering all the assumptions made by the programmer. Consequently, the designer must either accept on faith the program's results or check the results—experimentally, graphically, and numerically—in sufficient depth to satisfy himself that the programmer did not make dangerous assumptions or omit critical factors and that the program reflects fully the subtleties of the designer's own unique problem.

To underline the hazards of using a program written by somebody else, Petroski quotes a Canadian structural engineer on the use of commercial software:

> Because structural analysis and detailing programs are complex, the profession as a whole will use programs written by a few. Those few will come from the ranks of structural "analysts" . . . and not from the structural "designers." Generally speaking, their design and construction-site experience and background will tend to be limited. It is difficult to envision a mechanism for ensuring that the products of such a person will display the experience and intuition of a competent designer. . . . More than ever before, the challenge to the profession and to educators is to develop designers who will be able to

stand up to and reject or modify the results of a computer-aided analysis and design.[24]

The engineers who can "stand up to" a computer will be those who understand that software incorporates many assumptions that cannot be easily detected by its users but which affect the validity of the results. There are a thousand points of doubt in every complex computer program. Successful computer-aided design requires vigilance and the same visual knowledge and intuitive sense of fitness that successful designers have always depended upon when making critical design decisions.

Engineers need to be continually reminded that nearly all engineering failures result from faulty judgments rather than faulty calculations. For example, in the 1979 accident in the nuclear power plant at Three Mile Island, the level of the coolant in the reactor vessel was low because an automatic relief valve remained open, while for more than two hours after the accident began an indicator on the control panel said it was shut. The relief valve was opened by energizing a solenoid; it was closed by a simple spring. The designer who specified the controls and indicators on the control panel assumed that there would never be a problem of the valve's closing properly, so he chose to show on the panel not the valve position but merely whether the solenoid was "on" or "off." When the solenoid was "off," he assumed, the valve would be closed. The operators of the plant assumed, quite reasonably, that the indicator told them directly, not by inference, whether the valve was open or closed.[25] The choice made in this case may have seemed so simple and sensible as to be overlooked in whatever checking the design underwent. It might have been reexamined had the checker had experience with sticky relief valves or comprehension of the life-and-death importance of giving a nuclear power plant's operators direct and accurate information. This was not a failure of calculation but a failure of judgment.

A Cycle of Blunders

A cluster of newspaper articles that appeared in the first half of 1990 (a similar crop may be harvested in any half-year) has fattened my "failure" file folder and has led me to expect only more of the same under the accepted regimen of abstract, high-tech design. The magnitude of the errors of judgment in some of the reported failures (and in the numerous failures in Department of Defense projects that are protected from full public disclosure) suggests that engineers of the new breed have climbed to the tops of many bureaucratic ladders and are now making decisions that should be made by people with more common sense and experience.

The first oil spill of the year occurred in New York waters when a transfer line from the Exxon Bayway refinery to a tanker-loading dock on Staten Island spilled 500,000 gallons into Arthur Kill on the first day of January. Half a million gallons of oil will fill four storage tanks 30 feet in diameter and about 25 feet tall—the mid-size tanks one sees in a refinery tank farm. A few feet of a side seam had split in a section of pipe, and an automatic alarm valve, intended to shut off the flow, detected the leak but had been wedged by the operators so it would not shut. The valve had been wedged open for 12 years because the shut-down alarm was "too sensitive" and kept interrupting flow in the pipe line. In all those years, according to Exxon, the pipe had never leaked.[26]

On April 12, half a page of the *New York Times* was devoted to explaining how "smart" cars and highways would, in some indefinite future, "help unsnarl gridlock." A "major program of computerization and automation that would fundamentally alter the designs of vehicles and highways" is being actively promoted by the US Department of Transportation. Research is under way on "computerized dashboard maps" to tell the driver where he is, "roadside sensors and signals to manage the flow of traffic," and "sophisticated steering and speed controls that might someday even let cars drive themselves on specially instrumented highways."[27] Among the "sensors and signals" is a system intended to warn drivers of heavy traffic ahead and to inform them of possible alternative routes. Instruments will measure traffic flow and feed their information to a "control center." The control center will relay the information to a satellite; the satellite will send the information to cars "in the area." The information will then be displayed on a "dashboard monitor," which the driver can presumably study at leisure (perhaps viewing advertisements when not checking traffic conditions). Only

Rube Goldberg could have devised a more absurd (or expensive) scheme to do things in elaborate ways. A trial of the system will be conducted in Orlando, a most appropriate location for fantasy technology. The *Times* article made no mention of the $500 million already spent by the Department of Defense on a "smart truck" about a year earlier. That five-year program to develop an "autonomous truck" that could drive itself and find its way on and off highways had been phased out because of abysmally deficient performance. When the truck was being taught to guide itself on a highway, it could operate only at noon, with the sun directly overhead, because it was confused by shadows. Eventually, it was able to travel at 12 miles per hour on a straight, paved test track, and "to negotiate curves and to travel at any time of day and even at night using laser range finders." When it tried to make its way across open desert, "avoiding bushes and ditches along the way," its best performance was to guide itself about 600 yards at 2 miles per hour.[28]

On May 7, the *Wall Street Journal* gave a careful account of the expensive problems that poor design judgment and unreasonable production deadlines caused when General Electric introduced a new and insufficiently tested compressor in its domestic refrigerators in 1986. The new refrigerators featured rotary compressors rather than the reciprocating compressors that had been employed since the 1920s.[29]

Rotary compressors, already used in air conditioners, were attractive to GE managers because they were expected to be much cheaper to build. Many engineers learned in school a bit of folklore about the invariable superiority of rotating machinery over reciprocating machinery. Rotating gas compressors, however, require substantially more power than reciprocating compressors, and their high rotative speeds make them difficult to cool and lubricate.

The designers of the new compressors ignored the significant difference in performance requirements between air conditioners and refrigerators. In air conditioners, a convenient stream of air kept the body of the compressor, and thus the lubricating oil sealed inside, cool. Refrigerators lacked an equivalent air stream, and none was provided to cool the new compressors.

An experienced consultant told GE to buy compressors abroad or to learn how to make better compressors than those they planned to use. Although the designers had had no experience with rotary compressors, they rejected the advice and proceeded to develop a design that required tolerances smaller than any found in mass-produced machines of any kind. According to one of his former associates, the chief design engineer "figured you didn't need compressor-design experience to design a new compressor."[30]

The first of the new compressors were to be tested for the assumed lifetime of a refrigerator; however, the tests were cut short long before a "lifetime" had elapsed, and the misgivings voiced by the experienced technician who ran the tests were disregarded. This senior technician—who had worked in the testing lab for 30 years—reported that, although the compressors did not actually fail in the truncated testing program, "they didn't look right, either." Discoloration from high temperatures, bearing surfaces that looked worn, and a black oily crust on some parts pointed to eventual trouble with overheating, wear, and a breakdown of the sealed-in lubricating oil. The experience-based assessment was discounted because it came from a mere technician, a dirty-hands worker of much lower status than a scientific engineer.

The new refrigerators sold well, no doubt largely because they had other new features such as a "refreshment center" in the door. Troubles in the field did not begin until almost a year after their introduction. When the dimensions of the design debacle began to be clear, the company decided to forestall a likely customer revolt against all GE products by voluntarily replacing more than a million rotary compressors with reciprocating compressors at a cost of about $450 million.[31]

In May of 1990, NASA returned to the front pages with two blunders less chilling than the *Challenger* explosion but likely to waste hundreds of millions of dollars.

The Hubble space telescope, launched on April 24, had been confidently advertised as the answer to the problem of the atmosphere's interference with extremely faint light waves from far-distant heavenly bodies. The space telescope was expected to increase the diameter of the known universe by a factor of 7.[32] The first pictures were to be transmitted to earth a week after the launch. Seven weeks later, several unexpected happenings had postponed the first transmission to the end of the year, about 8 months behind schedule. Most significantly, an error had been made in grinding the large mirror, and it was impossible to bring any heavenly body into sharp focus. Computer experts proposed programs that would "enhance" the distorted images, but one may reasonably doubt the veracity of such enhancement.[33]

The first smaller-scale mishap involving Hubble occurred when the satellite carrying the telescope was launched from the shuttle vehicle. An electrical cable, connecting an adjustable antenna dish to the television transmitter, was kinked as it exited the shuttle, causing a significant reduction in the amount of antenna adjustment. Transmission to earth was interrupted by the inability of the antenna to be continuously pointed at the receiving station.

A few days later, newspaper readers learned that the telescope could not be pointed accurately at stars and planets. The controlling computer program had been based on "an outdated star chart," and therefore the telescope suffered a pointing error of about half a degree. Furthermore, the telescope developed a tendency to drift and to pick up other nearby stars just slightly brighter or dimmer than those it was intended to hold in focus.[34]

Finally, the severe vibration of the entire telescope satellite raised questions about its ability to obtain any information not available to ordinary telescopes located on the ground. An unanticipated (i.e., unthought of in the design) cycle of expansion and contraction of solar panel supports, as the spacecraft moved into and out of the earth's shadow, caused the panels to sway "like the slowly flapping wings of a great bird." The computer program for stabilizing the spacecraft, confused by the unexpected vibrations, called for stabilizing actions that exacerbated the vibration.[35]

Further deficiencies have turned up in Hubble's second year in orbit. Two gyroscopes (of six) have failed and two others exhibit signs of incipient failure. The Goddard high-resolution spectograph may have to be shut down because of intermittent loss of connection with its data computer. The flapping solar panels are attached to booms which have developed a jerky motion that may lead to their collapse and a catastrophic power loss. Although NASA hopes to send repair missions to Hubble in 1993, one wonders whether the repair missions will be able to keep ahead of the failures of one component after another.[36] These blunders resulted not from mistaken calculations but from the inability to visualize realistic conditions. They suggest that although a great deal of "hard thinking" may have been done to accomplish the stated missions of Hubble, the ability to imagine the mundane things that can go wrong remains sadly deficient at NASA.

While the Hubble telescope, a caricature of competent design, limped through the sky, the proposed space station *Freedom* entered the news. A review of the plans for that space station, to be assembled in space over a period of several years starting in 1995, revealed that the station might not be practicable because of excessive maintenance requirements outside the crew's living and working spaces. Astronauts would have to spend 2200 hours each year in space suits, outside the station, repairing and re-placing electronic gear, light bulbs, solar panels, batteries, and thermal blankets. The original estimate of maintenance time was 130 hours per year. (In 30 years of NASA space flights, astronauts have spent a total of 400 hours in "space walks.") An earnest former astronaut, testifying before a Congressional subcommittee, said that "the final assessment [of maintenance work] will be higher than I like," but concluded that the problem is not a "show-stopper."[37] John E. Pike of the Federation of American Scientists agreed with the astronaut that the need for so much maintenance is "politi-cally . . . not a death blow." "The station is too big to cancel," said Pike. "They'll muddle through. There's no other choice."[38] Unfortunately, that is probably an accurate assessment of the situation, as well as being a sad commentary on the nature of a bureaucratic organization endowed with un-limited money and extremely limited common sense.

Top-Down Design
Richard Feynman, the maverick physicist who served on the official panel reviewing the *Challenger* explosion, noted the inevitability of more failures and embarrassing surprises if NASA did not change radically the way its big projects were designed. He called the procedure being used "top-down design" and contrasted it with sensible "bottom-up" design that has been normal engineering practice for centuries.[39]

In bottom-up design, the components of a system are designed, tested, and if necessary modified before the design of the entire system has been set in concrete. In the top-down mode (invented by the military), the whole system is designed at once, but without resolving the many questions and conflicts that are normally ironed out in a bottom-up design. The whole system is then built before there is time for testing of components. The deficient and incompatible components must then be located (often a difficult problem in itself), redesigned, and rebuilt—an expensive and uncertain procedure.

Furthermore, as Feynman pointed out, the political problems faced by NASA encourage if not force it to "exaggerate" when explaining its reasons for needing large sums of money. It was "*apparently* necessary [in the case of the shuttle] to exaggerate: to exaggerate how economical the shuttle would be, to exaggerate how often it would fly, to exaggerate how safe it would be, to exaggerate the big scientific facts that would be discovered. 'The shuttle can make so-and-so many flights and it'll cost such-and-such; we went to the moon, so we can *do* it!'"[40]

Until the foolishness of top-down design has been dropped in a fit of common sense, the harrowing succession of flawed designs will continue to appear in high-tech, high-cost public projects.

The Unmet Need for Reflection

Although the need for the space station—whose estimated cost has risen from $8 billion in 1984 to more than $100 billion when operating costs are included (or $30–50 billion when operating costs are not included)—is purely political, some better assurance of a workable design might be gained by taking seriously a prescription for the design of pioneering projects made in the mid 1960s by a prominent British structural engineer. Sir Alfred Pugsley saw the need in such projects to give the chief engineer a "sparring partner," a senior engineer who was privy to essentially all the information available to the chief engineer and whose status was such that the chief could not ignore his comments and recommendations. This sparring partner would be given ample time to follow the design work and to study and think about

Figure 7.8
The Tacoma Narrows Bridge on November 7, 1940, just before the destruction of its suspended deck by a quartering wind with a velocity of 42 miles per hour.[71]

the implications of details as well as the "big" decisions made by the chief engineer.[41]

The hazards of permitting a chief engineer to determine all aspects of a complex project, without critical review, are less insidious and far-reaching than another hazard that Pugsley also warned against: the frequent adoption of a faulty doctrine by a whole profession. Pugsley's example of misplaced enthusiasm for a new doctrine was the collapse of the Tacoma Narrows suspension bridge in 1940, the "major lesson" of which was "the unwisdom of allowing a particular profession to become too inward looking and so screened from relevant knowledge growing up in other fields around it." Had the designers of the Tacoma Narrows Bridge known more of aerody-

Figure 7.9
The deck of the Tacoma Narrows Bridge disintegrating.

namics, he thought, the collapse might have been averted.[42] It is fairly certain, however, that if the relevance of aerodynamics to that design had been suggested by a person outside the network of "leading structural engineers," the advice would have been considered an attack on the profession of civil engineering. The experience of two engineers who published historical articles on the collapse of the bridge supports my surmise. The professional reaction to an article in *Engineering News-Record* by Dean James Kip French of the Columbia University engineering school prompted him to virtually retract its contents.[43] David Billington, an unorthodox professor of civil engineering at Princeton, was excoriated by several prominent bridge engineers when his paper on events leading to the collapse was published in a journal of the American Society of Civil Engineers.[44]

Billington, in a historical study of suspension bridges, argues convincingly that a design decision made in the 1920s by O. H. Ammann, designer of the George Washington Bridge in New York, "led directly to the failure of the Tacoma Narrows Bridge." Ammann decided that the deck of his bridge could be built without vertical stiffening and omitted the stiffening trusses that John Roebling and other suspension-bridge engineers had felt were necessary to keep winds from causing undulation of the bridge deck. Ammann's reasoning appealed to many in the civil engineering profession, and several long, slender, and disturbingly flexible suspension bridges were built in the 1930s (including the Golden Gate Bridge, which was stiffened after a harrowing experience with crosswinds in 1951).[45]

After the Tacoma Narrows Bridge fell, structural engineers found that a sense of history might have tempered their enthusiastic acceptance of Ammann's design precept. They learned, as Billington points out, that published records of suspension bridges in Europe and America "described nineteenth-century failures that were amazingly similar to what they saw in the motion pictures of the Tacoma collapse."[46]

Billington's article was characteristically greeted by engineers as an "attack upon the leading figures of the period and especially upon O. H. Ammann." Rebuttal was necessary, according to Billington's many critics, in order to "remove the undeserved blame" leveled at several bridge designers and to "preserve their proper position in the history of engineering."[47]

The need to justify the way engineers do things is unfortunately present even when ill-considered systems lead operators to make fatally wrong judgments.

The missile cruiser USS *Vincennes* was equipped with a billion-dollar "state-of-the-art" air defense system called Aegis. On July 3, 1988, the ship shot down an Iranian civilian airliner and thus killed 300 people.[48] The Aegis system had received IFF (Identification of Friend or Foe) signals for both military and civilian planes, yet the ship's radar indicated only one plane, and the decision was made to destroy it. (No radar or any other existing equipment will identify a plane by its physical shape and size alone.) Later the Navy decided that an enlisted man had misinterpreted the signals on his visual display and that therefore the captain was not at fault for ordering the destruction of the civilian airplane.

As with most "operator errors" that have led to major disasters, the operators aboard the *Vincennes* had been deluged with more information than they could assimilate in the few seconds before a crucial decision had to be made. It is a gross insult to the operators who have to deal with such monstrous systems to say, as the Navy did, that the Aegis system worked perfectly and that the tragedy was due to "operator error."[49]

The designers of Aegis, which is the prototype system for SDI, grossly underestimated the demands that their designs would place on the operators—who often lack knowledge of the idiosyncrasies and limitations built into the system. Disastrous errors of judgment are inevitable so long as operator error rather than designer error is routinely considered as the cause of disasters. Hubris and an absence of common sense in the design process set the conditions that produce the confusingly overcomplicated tasks that the equipment demands of operators. Human abilities and limitations need to be designed into systems, not designed out.

If we are to avoid calamitous design errors as well as those that are merely irritating or expensive, it is necessary that engineers understand that such errors are not errors of mathematics or calculation but errors of engineering judgment—judgment that is not reducible to engineering science or to mathematics.

Here, indeed, is the crux of all arguments about the nature of the education that an engineer requires. Necessary as the analytical tools of science and mathematics most certainly are, more important is the development in student and neophyte engineers of sound judgment and an intuitive sense of fitness and adequacy.

No matter how vigorously a "science" of design may be pushed, the successful design of real things in a contingent world will always be based more on art than on science. Unquantifiable judgments and choices are the elements that determine the way a design comes together. Engineering design is simply that kind of process. It always has been; it always will be.

Notes to the Text

Chapter 1

1. Edwin T. Layton, Jr., "American Ideologies of Science and Engineering," *Technology and Culture* 17, no. 4 (October 1976), pp. 688–701; quotation from p. 696.

2. "Report on Engineering Design," *Journal of Engineering Education* 51, no. 8 (April 1961), pp. 645–660.

3. Richard S. Kirby et al. (*Engineering in History* [New York, 1956 and 1990], p. 2) find that this definition, and the many others put forth by lexicographers and even by engineering organizations since the 1880s, "no longer seem adequate." Predictably, Kirby and his engineer associates add "scientific" to the attributes of engineering, but their definition retains the original three words: "the art of."

4. José Ortega y Gasset, in his essay "Man the Technician" in his *History as a System* (New York, 1962), observes: "Food, heat, etc. . . . are human necessities because they are objective conditions of life understood as mere existence in the world [but] man has no desire to 'be in the world'; he wants to live well. Man is the only animal that considers necessary . . . the objectively superfluous."

5. Homer E. Newell, *Beyond the Atmosphere: Early Years of Space Science* (Washington, D.C., 1980), p. 430; John M. Logsdon, *The Decision to Go to the Moon* (Chicago, 1970), pp. 106–110, 127.

6. David R. Smith, "They're Following Our Script: Walt Disney's Trip to Tomorrowland," *Future*, no. 2 (May 1978), pp. 54–63.

7. Henry J. Kauffman, *American Axes* (Brattleboro, Vt., 1972), p. 23; Vernard Foley and Richard H. Moyer, "The American Axe: Was It Better?" *Chronicle* (Early American Industries Association) 30 (June 1977), pp. 28–32. The American axe became a more effective woodsman's tool than the European colonial trade axe, from which it developed. The heavier head reduced total effort by requiring a slower swing for the same blow. The heavy poll moved the center of percussion to the center of the handle, thus reducing the "sting" to the hands; moving the center of balance reduced the effort of "aiming" the blade during the swing. The axe's evolution proceeded with continual feedback, which produced many variations on a heavily polled axe with a relatively broad, short bit. See also V. Foley et al.,

196

"Dynamics of the Axe Swing," *Ergonomics* 21, no. 11 (1978), pp. 925–930, and "Axe Use and Efficiency," ibid. 24, no. 2 (1981), pp. 103–109. The Warren Axe and Tool Company was in Warren, Pa. Its 1916 trade catalogue, from which figure 1.3 is taken, is in Hagley Museum and Library at Wilmington, Delaware. Another American axe maker's catalogue (1907) in the Hagley Museum lists 65 patterns, each available in various weights (generally from 3 to 6 pounds, in quarter-pound increments).

8.   Charles Haines suggested this incident and supplied the proper jargon.

9.   Frank D. Prager and Gustina Scaglia, *Brunelleschi: Studies of His Technology and Inventions* (Cambridge, Mass., 1970), pp. 85–88; Gustina Scaglia, "Building the Cathedral in Florence," *Scientific American* 264 (January 1991), pp. 66–72, esp. 69.

10. A basic difficulty in developing a science of design was voiced by an English user of computer models of turbulent flow around aircraft, which some researchers hoped would eliminate the need for most wind-tunnel testing: "Computers merely rearrange information; only experiments can generate it." (Peter Bradshaw, letter to editor, *Astronautics and Aeronautics* 13, no. 9 [September 1975], p. 6.) H. M. Collins, in his article "Expert Systems and the Science of Knowledge," in *The Social Construction of Technological Systems,* ed. W. E. Bijker et al. (Cambridge, Mass., 1989), concludes, quite reasonably, that "the design of crafted artifacts . . . is not an exact science; in addition to formal theories, principles of design rest on traditional ways of doing things—an artisan's skills and bodies of tacit knowledge that cannot be fully articulated" (p. 331), and that a person who uses a computer program in design must supply "that part of the iceberg of cultural [i.e., tacit] knowledge that cannot be programmed" (p. 345).

Tacit and cultural knowledge are the bases for common sense, and, as Douglas Hofstadter points out, "there is no program that has common sense; no program that learns things it has not been explicitly taught how to learn; no program that can recover gracefully from its own errors" (quoted in Hubert and Stuart Dreyfus, *Mind Over Machine: The Power of Human Intuition and Expertise in the Era of the Computer* [New York, 1986], p. 67).

11. Walter G. Vincenti, *What Engineers Know and How They Know It: Analytical Studies from Aeronautical Engineering* (Baltimore, 1990). In this pioneering book, Vincenti describes the intellectual basis of engineering design with a generality that goes far beyond his aeronautical engineering examples.

12. See Charles H. Gibbs-Smith, *Aviation: A Historical Survey* (London, 1970),

pp. 58–59, 96–97, 110, 126–127, 139. Gibbs-Smith argues that failure to grasp the importance of three-dimensional control delayed the development of airplanes in Europe until 1909.

Folklore credits the Wrights' use of wing warping to their observation of turkey buzzards in flight. Fred C. Kelly, in *The Wright Brothers* (New York [paperbound], 1956), p. 47, corrects the misconception and quotes a reflection by Orville Wright: "Though the Wrights often studied the flight of birds in the hope of learning something, they did not at first learn anything of use to them in that way. After they had thought out certain principles, they then watched the birds to see if they used the same principles. As Orville Wright wrote long afterward, 'Learning the secret of flight from a bird was a good deal like learning the secret of magic from a magician. After you once know the trick and know what to look for, you see things that you did not notice when you did not know exactly what to look for.'"

13. Walter G. Vincenti, "How Did It Become 'Obvious' That an Airplane Should Be Inherently Stable?" *American Heritage of Invention and Technology* 4, no. 1 (spring-summer 1988), pp. 50–56. See also chapter 3 of Vincenti, *What Engineers Know* (n. 11 above).

14. Nevil Shute, in *Slide Rule: The Autobiography of an Engineer* (New York [paperbound], 1964), p. 47, tells of his experience as a designer with aircraft "stability." In the 1920s, when he was working for Geoffrey de Havilland, a pioneer builder of airplanes, de Havilland took him up in his personal plane one day "to show me what he meant when he talked at my desk about 'stability.'" Compare James Gleick's comment (*Chaos: Making a New Science* [New York, 1987], p. 262) that "you don't see something until you have the right metaphor to let you perceive it."

15. Edwin T. Layton, Jr., "Mirror-Image Twins: The Communities of Science and Technology in 19th-Century America," *Technology and Culture* 12, no. 4 (October 1971), pp. 562–580.

16. Otto Mayr (*Technology and Culture* 17, no. 4 [October 1976], p. 662), after identifying several abstract concepts, such as "control schemes (feedback and program control), oscillators (escapement, hydraulic ram, electrical doorbell, electronic oscillator)," adds: "Such concepts have their origins within technology, and have distinct histories and genealogies of their own. I am always amazed how this aspect of technology is neglected by historians."

17. James Gleick observes: "Simulations break reality into chunks, as many as possible but always too few. A computer model is just a set of arbitrary rules, chosen by

programmers. . . . Whenever a good physicist examines a simulation, he must wonder what bit of reality was left out, what potential surprise was sidestepped. Libchaber [a physicist] liked to say that he would not want to fly in a simulated airplane—he would wonder what had been missed." (*Chaos*, p. 210) Edwin T. Layton, Jr., in "Millwrights and Engineers, Science, Social Roles, and the Evolution of Turbines in America," in *The Dynamics of Science and Technology*, ed. W. Krohn et al. (Dordrecht, 1978), explains (p. 78) how "idealization involved assumptions based on mathematical necessity rather than physical reality."

18. *Time*, May 14, 1945, p. 3. I am indebted to Bruce E. Seely for this reference.

19. Carl Mitcham, "Types of Technology," *Research in Philosophy and Technology* 1 (1978), pp. 229–294; quotation from p. 244.

20. H. W. Dickinson's *Short History of the Steam Engine* (Cambridge, 1938) has not been superseded, although specialist literature has burgeoned. I told part of the story in "The Origins of the Steam Engine," *Scientific American* 210, no. 1 (January 1964), pp. 98–107.

21. Mark Clark, The Design of Orbital Docking Mechanisms at NASA: A History, Master's thesis, University of Houston, 1987, pp. 50, 68–77.

22. Ibid., p. 87.

23. John Van der Zee, *Gate: The True Story of the Design and Construction of the Golden Gate Bridge* (New York, 1986); Stephen Cassady, *Spanning the Gate* (Mill Valley, Calif., 1979).

24. George Basalla, *The Evolution of Technology* (Cambridge, 1988), pp. 21, 26, 208–209.

25. David Pye, *The Nature and Aesthetics of Design* (New York, 1978), p. 65.

26. On American artists, see Edward C. Carter II et al., eds., *Latrobe's View of America, 1795–1820: Selections from the Watercolors and Sketches* (New Haven, 1985); on Morse, see Paul J. Staiti, *Samuel F. B. Morse* (Cambridge and New York, 1989) and Brooke Hindle, *Emulation and Invention* (New York, 1981); on Fulton, see Hindle. It should be noted also that many prominent engineers of the nineteenth century were superb draftsmen.

The organization of American technology in the first half of the nineteenth century tended naturally to follow the pattern set by the world of art. American societies for the promotion of the mechanical arts were influenced heavily by the London Society of Arts, which was concerned with both the fine arts and the mechanical arts. The Franklin Institute, organized in 1824, awarded prizes at its annual exhibitions and sponsored classes in mechanical and architectural drawing. The American Institute of the City of New York held annual industrial fairs and, like the London Society of Arts, maintained a permanent exhibition of machines and models for the observation and study of mechanics and inventors. See Brooke Hindle, "The Underside of the Learned Society in New York, 1754–1854," in *The Pursuit of Knowledge in the Early American Republic,* ed. A. Oleson and S. C. Brown (Baltimore, 1976).

27. David P. Billington, *The Tower and the Bridge* (New York, 1983), pp. 4, 155.

28. David P. Billington, *Robert Maillart's Bridges: The Art of Engineering* (Princeton, 1979), pp. 21–26, 34–41. The quotation is from Billington's book *The Tower and the Bridge* (New York, 1983), p. 158. David Billington has also published two editions of *Thin Shell Concrete Structures* (New York, 1965 and 1982), a reference book on design and construction in which he emphasizes the fact that successful designers of such structures are thoroughly familiar with the *performance* of thin-shell structures already built and also have a deep understanding of the *methods and problems of construction* of such structures.

29. Reese V. Jenkins, "Elements of Style: Continuities in Edison's Thinking," in *Bridge to the Future: A Centennial Celebration of the Brooklyn Bridge,* ed. M. Latimer et al. (New York, 1984). On sources of Edison's elements of style (an inspired use by Jenkins of E. B. White's phrase), see chapter 5 below.

   See other papers on design in the Brooklyn Bridge volume, particularly those of Robert M. Vogel and Edwin T. Layton, Jr.

30. "Elements of Style" (n. 29 above).

31. Institution of Civil Engineers, *Proceedings of a Joint Conference [of I.C.E., A.S.C.E., Canadian Soc. for C.E.] held in Torquay 20–25 September 1981* (London, 1982), pp. 135–136.

32. "Deep knowledge" means to me comprehensive understanding and appreciation of the many, many facets of a situation. Herbert and Stuart Dreyfus, in their book *Mind over Machine* (New York, 1986), call it "expertise" and give many examples

of the intuitive, unreflective, expert performances of baseball players, x-ray diag-nosticians, business managers, and so on (pp. 30–35). On the other hand, Michael Rychener, in proposing "expert systems" for design, defines deep knowledge as "knowledge at a deep theoretical level, as opposed to experiential, superficial knowledge" that "is often expressed as mathematically rigorous models, as opposed to heuristic rules" (M. Rychener, ed., *Expert Systems for Engineering Design* [Boston, 1988], pp. 22–23). Rychener's definition, reflecting an academic view of design, epitomizes the gulf that will always exist between expert systems and experts—see "expert systems" in the index of the Dreyfus book.

33. Henry S. F. Cooper, *Thirteen: The Flight that Failed* (New York, 1973), p. 114 and passim. The entire book is a study of the performance of engineers and their structures and systems. Many significant insights into the engineering mind will reward the reader of this exciting, well-told tale. See also Charles R. Pellegrino and Joshua Stoff, *Chariots for Apollo* (New York, 1985). The landing module became a lifeboat for the command module, a situation unimagined except in a 1960 design consultation (mentioned on p. 190).

34. Louis L. Bucciarelli, "Engineering Design Process," in *Making Time: Ethnographies of High-Technology Organizations,* ed. F. Dubinskas (Philadelphia, 1988), esp. pp. 96–97.

35. James W. Althouse III, letter to author, February 10, 1989.

36. C. E. Evanson and H. A. Edmonds, "Blackboard Drafting," *Machine Design* 34 (July 19, 1962), pp. 148–151. The PDT system is described in this article. The TAB Engineers hold a "one-day training session" for management and engineering supervisors to ensure that they are "thoroughly sold on the benefits of PDT." The client's engineers and draftsmen are then trained in two-week courses. At least eight articles describing PDT appeared in technical journals in 1962 (as listed in the *Applied Science and Technology Index*).

37. *Mechanical Engineering* 87 (May 1965) is an issue devoted to computer-aided design. Two articles were written by IBM research engineers; an article on the MIT "CAD Project" was written by Robert W. Mann.

38. Bucciarelli (n. 34 above, p. 96) notes an uneasiness among designers in a group when the block diagrams do not describe reality.

39. The advent of computers has not eliminated the need for decisions. In describing

Boeing's experience with the "paperless" (i.e., fully computerized) design of its 777 widebody airplane, John Holusha points out that thus far "the basic outlines of the craft have been decided, but now engineers have to make millions of detailed decisions so the design can be largely finished by next March" (*New York Times*, November 10, 1991).

40. David P. Billington, *Thin Shell Concrete Structures*, second edition (New York, 1982), p. xii.

Chapter 2

1. Richard P. Feynman, *"What Do You Care What Other People Think?"* (New York, 1988), p. 54.

2. A. Rupert Hall, "Guido's *Texaurus*, 1335," in *On Pre-Modern Technology and Science*, ed. B. Hall and D. C. West (Malibu, 1976), pp. 24, 43. The description and illustration are of a portable military bridge whose components were small enough to be carried by horses. In the 1989 *Oxford English Dictionary*, the earliest citations of "mind's eye" are from Hoccleve (1412) and from Hamlet's vision of his dead father (1602); s.v. *Eye*.

3. Rudolf Arnheim, in his book *Visual Thinking* (Berkeley, 1969), develops this proposition and, in his final chapter, develops an impassioned argument for visual thinking and not merely visual aids and visual entertainment in public education—a much more perceptive statement than any other I have seen.

4. Maya Pines, "We Are Left-Brained or Right-Brained," *New York Times Magazine*, September 9, 1973, p. 32 ff., esp. p. 33.

5. Stanley L. Englebardt, "Are You Thinking Right?" *Readers Digest* 188 (February 1988), pp. 41–45.

6. Roger Sperry, "Some Effects of Disconnecting the Cerebral Hemispheres," *Science* 217 (September 24, 1982), pp. 1223–1226. This article is Sperry's 1981 Nobel Prize lecture. See also a biographical sketch of Sperry in the announcement of the Nobel Prize in *Science* 214 (October 30, 1981), pp. 517–518.

7. Maya Pines, *The Brain Changers* (New York, 1973), p. 146.

8. Sperry (n. 6 above), p. 1224.

9. Robert R. Holt, "Imagery: The Return of the Ostracized," *American Psychologist* 19, no. 4 (April 1964), pp. 254–264; Ned Block, ed., *Imagery* (Cambridge, Mass., 1981), p. 1. See also Geir Kaufmann, *Visual Imagery and Its Relation to Problem Solving* (Bergen, 1979).

10. Francis Galton, *Inquiries into Human Faculty and Its Development* (London, 1883 and 1907), pp. 58–60, 78; Ann Roe, "A Study of Imagery in Research Scientists," *Journal of Personality* 19 (1950–51), pp. 495–470. Roe notes (p. 461) Galton's argument for "further development and utilization of visual imagery."

11. Galton, *Inquiries* (n. 10 above).

12. Robert Kargon, "Model and Analogy in Victorian Science," *Journal of the History of Ideas* 30, no. 3 (July–September 1969), pp. 423–436; Richard Olson, *Scottish Philosophy and British Physics 1750–1880* (Princeton, 1975), pp. 3–8.

13. Pierre Duhem, *The Aim and Structure of Physical Theory,* tr. P. P. Weiner (New York, 1962), pp. 64, 70, 71.

14. William James, *The Essential Writings,* ed. B. W. Wilshire (New York, 1971), pp. 68, 69.

15. Arthur I. Miller (*Imagery in Scientific Thought: Creating 20th-Century Physics* [Cambridge, Mass., 1986]) argues that "thinking in images is an essential ingredient of scientific research of the highest creativity" (p. 222). For an extended essay on the significance of Einstein's nonverbal thinking, see Gerald Holton, "On Trying to Understand Scientific Genius," *American Scholar* 41, no. 1 (winter 1971–72), pp. 95–110, reprinted in Holton's *Thematic Origins of Scientific Thought* (revised edition, Cambridge, Mass., 1988). On Einstein's difficulty in translating thoughts into words, see Pines (n. 7 above), pp. 158–159.

16. Freeman Dyson, *Disturbing the Universe* (New York, 1979), p. 62. Dyson tells in this book (pp. 53–76) how he became the interpreter to other physicists of Feynman's visual approach to quantum mechanics. See also Miller, *Imagery* (n. 15 above), p. 170.

17. Robert S. Root-Bernstein, "Visual Thinking: The Art of Imagining Reality," *Trans-*

actions (American Philosophical Society) 75, part 6 (1985), pp. 50–67. The author wonders (p. 63) "to what extent visual imagination is dependent or independent of the kinesthetic skills of drawing or modeling." I believe that a person who has those drawing skills can use intelligently a computer programmed to supply ready-made geometric figures, whereas one who has not learned to draw will help to create, in the words of Gary R. Bertoline, "more visual garbage than the world has ever seen" ("The Role of Computers in the Design Process," *Engineering Design Graphics Journal* 52, no. 2 [Spring 1988], pp. 18–22, 30). Root-Bernstein opens his paper with a tribute to Brooke Hindle and his book *Emulation and Invention* (New York, 1981).

18. Agnes Arber, *The Mind and the Eye* (Cambridge, 1954), p. 5; Pines (n. 4 above), p. 32.

19. Eugene S. Ferguson, *Oliver Evans: Inventive Genius of the American Industrial Revolution* (Greenville, Del., 1980), p. 20.

20. Quoted in Eugene S. Ferguson, "Kinematics of Mechanisms from the Time of Watt," *Bulletin* (U.S. National Museum) 228 (1963), pp. 185–230; quotation on p. 195.

21. Richard Beamish, *Memoir of the Life of Sir Marc Isambard Brunel* (London, 1862), p. 321. James Nasmyth's *Autobiography,* ed. Samuel Smiles (London, 1883), p. 125, echoes Brunel: "Mechanical drawing is the alphabet of the engineer. Without this alphabet the workman is merely 'a hand.'" See Christopher Polhem's mechanical alphabet in chapter 5 below.

22. Nasmyth (n. 21 above), p. 272. Nicola Tesla (1856–1943) built imaginary electrical machines in his mind and ran them "for weeks on end . . . periodically checking them for signs of wear" (Curt Wohleber, "The Work of the World," *American Heritage of Invention and Technology* 7, no. 3 [winter 1992], pp. 44–52).

23. Walter P. Chrysler and Boyden Sparks, *Life of an American Workman* (New York, 1950), p. 43. I am indebted to John Rumm for this reference.

24. Nasmyth (n. 21 above), p. 97.

25. Quoted in *Dictionary of American Biography* (New York, 1928–1936) s.v. Peter Cooper Hewitt (1861–1921).

26. Elmer Sperry, "The Spirit of Invention in an Industrial Civilization," in *Toward Civilization,* ed. C. A. Beard (New York, 1930); quotation on pp. 63–64.

27. Thomas Parke Hughes, *Elmer Sperry, Inventor and Engineer* (Baltimore, 1971), pp. 51–52.

28. James R. Hansen, *Engineer in Charge: A History of the Langley Aeronautical Laboratory, 1917–1958* (Washington, D.C., 1987), pp. 311–341.

29. Ibid., pp. 312, 327–328, 454.

30. Ibid., pp. 332–336.

31. *Virginian Pilot and Ledger Star* (Norfolk), May 18, 1985, pp. A1, A3.

32. For example, Alex Roland, *Model Research: The National Advisory Committee for Aeronautics, 1915–1958.* NASA SP-4103 (Washington, 1985), pp. 280–281.

33. Hansen (n. 28 above), p. 341.

34. Charles R. Pellegrino and Joshua Stoff, *Chariots for Apollo: The Making of the Lunar Module* (New York, 1975), pp. 185–186.

35. Robert Pirsig, *Zen and the Art of Motorcycle Maintenance* (New York, 1974), p. 311.

36. Dave McCormick was the engineer who taught me to say "let's go see." I was then a junior engineer in a large chemical plant.

    The construction engineer or works engineer who spends his day in the office is much like the medical doctor who makes a diagnosis mostly on the basis of laboratory tests and pays only cursory attention to the patient before him. Dr. William A. Dock, a distinguished teacher and a leading diagnostician, taught his students the cardinal rule of medicine: "Go to the patient, because that is where the diagnosis is." The wording came from Sutton's Law, named after the bank robber who robbed banks "because that's where the money is." A good clinical examination and a thorough history were important, Dock said, but "it was paramount to talk with the patient with focused, painstaking attention to detail because therein lay the key to the diagnosis." See Lawrence Reed, M.D., letter to the editor, *New York Times,* November 6, 1990, and Dr. Dock's obituary notice, *ibid.,* October 23, 1990.

37. A brainstorming session in the Navy Department in Washington was reported in the *New York Times* on May 20, 1956 ("Federal Brains Brace for Storm") and on May 24, 1956. Osborn's textbook, *Applied Imagination* (New York, 1953), went through a "third revised edition" in 1963, when 120,000 copies were in print. A summary of creativity courses and activities is given in the foreword of the third edition. Osborn was a prime mover in establishing the Creative Education Foundation, based at State University of New York at Buffalo, which has published the *Journal of Creative Behavior* since 1967 and which has sponsored an annual week-long "Creative Problem-Solving Institute."

38. Selected papers from the first three "University of Utah Conferences: The Identification of Creative Scientific Talent, Supported by the National Science Foundation" were published in *Scientific Creativity: Its Recognition and Development,* ed. C. W. Taylor and F. Barron (New York, 1963). The fourth conference was reported in *Creativity: Progress and Potential* (New York, 1964). The "most pressing" research problem was still "the determination of criteria of creativity" (ibid., p. 156; see also pp. 9, 178).

   Fads fade, but some seem to come and go in cycles. Here is a 1991 reincarnation: In its Wilmington corporate headquarters, E. I. du Pont de Nemours & Co. has recently installed a new Center for Creativity and Innovation. Promoted by its director, a 62-year-old former technical director of industrial fibers, the "creativity" activities are extensive and are prospering, according to a local newspaper reporter, "because the corporate giant that made safety a catch phrase is now making a similar push for creativity." The director of the center is quoted as saying that "probably the biggest obstacle we have is that you say the word creativity and some people think someone is doing something silly." (Gary Soulsman, "Du Pont's Push for Creativity: It's the New Corporate Imperative, So Turn Your Minds Loose," *Sunday News-Journal* [Wilmington], June 16, 1991)

39. See chapters 6 and 7 below.

40. One of the *Wall Street Journal*'s front-page feature stories (August 3, 1987) was entitled "Pivotal People: In the Computer Age, Certain Workers Are Still Vital to Success." That article described some of the skills and "knacks" of a blast furnace foreman, a garment maker's "pattern cutter," a brewmaster at a small brewery, and a gas well driller. The pivotal people are experts. Their indispensable knowledge comes from experience, from noticing and watching, and from their senses of timing and sequence. They use their expertise unselfconsciously and intuitively.

41. John R. Harris, Industry and Technology in the Eighteenth Century: Britain and

France, inaugural lecture, University of Birmingham [England], May 1971. Another statement of Harris' thesis is in his "Skills, Coal and British Industry in the Eighteenth Century," *History* 61 (June 1976), pp. 167–182; a brief, illustrated statement is in his "Rise of Coal Technology," *Scientific American* 231 (August 1984), pp. 92–97.

42. John R. Harris, "The Diffusion of English Metallurgical Methods to Eighteenth-Century France," *French History* 2, no. 1 (1988), pp. 22–44; quotations on p. 40.

43. The "puddling" of iron was a manual and mental skill of the nineteenth and the early twentieth centuries that involved strength and subtle judgment. A French author, Jean-Paul Courtheoux, explained the difference between knowing what the puddler did and knowing how to do it in the phrase "knowledge of the process, ignorance of the craft," quoted in Harris, "Skills" (n. 41 above), p. 178. I believe that an engineer who appreciates the distinction between process and craft will take care to review new design ideas with those who practice the craft.

Chapter 3

1. José Ortega y Gasset, "Man the Technician," in his *History as a System* (New York, 1961); quotations on pp. 151, 155. In a brief *tour de force,* Robert P. Multhauf recently traced the terminology identifying practitioners of the useful arts from the time of Plato to the middle of the nineteenth century. The first to be called engineers were Leonardo, Francesco di Giorgio, and their contemporaries. Their first love was military engineering, followed by the "flamboyant projects" illustrated in Renaissance machine books. Of nineteenth-century British engineers, Multhauf observes: "their inventions . . . seem to be directed at the mundane needs of society rather than the flamboyant projects characteristic of governments." See Multhauf, "Some Observations on the Historiography of the Industrial Revolution," in *In Context: History and the History of Technology,* ed. S. H. Cutcliffe and R. C. Post (Bethlehem, Pa., 1989).

2. Lynn White, Jr., "The Flavor of Early Renaissance Technology," in *Developments in the Early Renaissance,* ed. B. A. Levy (Albany, 1972); quotation on p. 41.

3. A 1980 study of the employment patterns of 22,000 members of the Swedish Society of Graduate Engineers found (a) 72% working with existing technologies, (b) 10% engaged in training others to maintain our existing technological world,

and (c) 18% developing new technologies. See Per Jacobsson, "Humaniora och teknik—kan de mötas? Om humanioras och samhällsvetenskapernas roll i civilingenjörsutbildningen," *UHÄ-rapport* 6 (1982), p. 73.

4. This theme was developed a bit further in my "La fondation des machines moderne: des dessins," *Culture technique* [Neuilly-sur-Seine-Paris], no. 14 (June 1985), pp. 182–207. The article was translated from my English typescript "Paper Foundations of Modern Machines."

5. See Bertrand Gille, *Renaissance Engineers* (London, 1966); Ladislao Reti, "Francesco di Giorgio Martini's Treatise on Engineering and Its Plagiarists," *Technology and Culture* 4 (summer 1963), pp. 287–298. Gustina Scaglia has published facsimile editions of Mariano Taccola's notebooks *De Machinis* (Wiesbaden, 1971) and *De Ingeneis* (Wiesbaden, 1984), with texts in English. She published a monumental "Typology of Leonardo da Vinci's Machine Drawings and Sketches," in *Leonardo da Vinci, Engineer and Architect,* ed. P. Galluzzi and J. Guillaume (Montreal Museum of Fine Arts, 1987), and a definitive study of Francesco di Giorgio's notebooks: *Francesco di Giorgio. Checklist and History of Manuscripts and Drawings in Autographs and Copies from ca. 1470 to 1687 and Renewed Copies (1764–1839)* (Bethlehem, Pa., 1992).

6. Gustina Scaglia, "Drawings of Machines for Architecture from the Early Quattrocento in Italy," *Journal* (Society of Architectural Historians) 25, no. 2 (May 1966), pp. 90–114. For Taccola's influence on later engineers, see Scaglia's edition of Taccola's *De Ingeneis,* vol. 1, pp. 28–31.

7. For Greek and Roman war machines, see A. G. Drachmann, *The Mechanical Technology of Greek and Roman Antiquity* (Copenhagen and Madison, Wisconsin, 1963). War machines, pumps, and water mills were touched upon in the Renaissance books called "theaters of machines." For example, see Agostino Ramelli, *The Various and Ingenious Machines of Agostino Ramelli,* tr. M. Teach Gnudi (London and Baltimore, 1976; New York, 1987).

8. Christopher Duffy, *Siege Warfare* (London, 1979), pp. 8–22. F. L. Taylor (*The Art of War in Italy 1494–1529* [Cambridge, 1921]) emphasizes the effectiveness of the new French artillery and, within a very few years, the strong defensive responses of Italian engineers. For a detailed study of the economic and human realities of the new warfare, see Simon Pepper and Nicholas Adams, *Firearms and Fortifications: Military Architecture and Siege Warfare in 16th-Century Siena* (Chicago, 1986).

9. Duffy (n. 8 above), pp. 23–42; J. R. Hale, "Early Development of the Bastion, 1450–1534," in *Europe in the Late Middle Ages,* ed. J. R. Hale (Evanston, Ill., 1965).

10. Duffy (n. 8 above), p. 41; Horst de La Croix, "Military Architecture and the Radial City Plan in Sixteenth Century Italy," *Art Bulletin* 42, no. 4 (December 1960), pp. 261–290, esp. pp. 273–274.

11. de la Croix (n. 10 above), pp. 274–275, 281–283, and 283 n. 86; Duffy (n. 8 above), p. 250; Dennis H. Mahan, *A Treatise on Field Fortification,* third edition (New York, 1860), p. 153.

12. Laurio Martines writes of patronage in his *Power and Imagination: City-States in Renaissance Italy* (New York, 1979). The patrons made use of the imagination of those patronized in order to maintain their power.

13. Ladislao Reti, ed., *The Unknown Leonardo* (New York, 1974), pp. 6–7.

14. For a biographical sketch of Vauban by Henry Guerlac, see *Dictionary of Scientific Biography* (New York, 1970–80), vol. 13, pp. 590–595. The lasting fame of Vauban was brought home to me in Paris in 1983; that year was "Vauban Year," marking the 250th anniversary of his birth. Dozens of celebrations and exhibitions were held in all parts of France throughout the year.

15. Frederick B. Artz, *The Development of Technical Education in France 1500–1800* (Cambridge, Mass., 1966), pp. 87–101.

16. Ibid., pp. 81–86.

17. A fresh analysis of the Ecole Polytechnique and its influence on the US Military Academy is Peter M. Molloy's Ph.D. dissertation Technical Education and the Young Republic: West Point as America's Ecole Polytechnique, 1802–1833 (Brown University, 1975). A significant article on the engineering school at Mézières by René Taton may be found in *Enseignment et diffusion des sciences en France au XVIIIᵉ siècle,* ed. R. Taton (Paris, 1964).

18. According to Molloy (n. 17 above), Thayer adopted the Ecole Polytechnique's curriculum of 1795–1804—a period when that "genuine engineering school" was expected to supply both military and civil engineers. The heavily scientific curriculum, introduced at the Ecole Polytechnique after 1815, was not generally adopted

by American engineering schools. See also George S. Emmerson's *Engineering Education: A Social History* (Newton Abbot and New York, 1973), a history of wide scope that outlines the social realities and intellectual underpinnings of engineering schools in Europe and North America.

Chapter 4

1. William M. Ivins, Jr., *Prints and Visual Communications* (Cambridge, Mass., 1953, 1969), p. 160. In his book *On the Rationalization of Sight* (New York, 1975), Ivins observes: ". . . sight has today become the principal avenue of the sensuous awareness upon which systematic thought about nature is based. Science and technology have advanced in more than direct ratio to the ability of men to contrive methods by which phenomena which otherwise could be known only through the senses of touch, hearing, taste, and smell have been brought within the range of visual recognition and measurement and thus become subject to that logical symbolization without which rational thought and analysis are impossible." (p. 13) Ivins saw the rationalization of sight as "the most important event of the Renaissance."

    Samuel Y. Edgerton, Jr., an architectural historian, has supplied "hard data" to support Ivins' insight. See Edgerton's articles "Linear Perspective and the Western Mind: The Origins of Objective Representation in Art and Science," *Cultures* (UNESCO Press) 3, no. 3 (1976), pp. 77–104, and "The Renaissance Artist as Quantifier," in *The Perception of Pictures,* vol. 1, ed. M. A. Hazen (New York, 1980). Ivins demonstrates a symbiotic relationship, surprising to most of us, between linear perspective and Western science.

2. Paolo Rossi, *Philosophy, Technology, and the Arts in the Early Modern Era,* tr. S. Attanasio (New York, 1970), pp. 28–29. The omitted phrase is "(as Erwin Panofsky has pointed out)."

3. Elizabeth L. Eisenstein, *The Printing Press as an Agent of Change* (Cambridge, 1979), volume 1, p. 290. The quoted words are those of James Ussher, Archbishop of Armagh (1581–1656). Pliny wrote of "the manifold hazards in the accuracy of copyists" in attempting to illustrate a catalogue of medicinal herbs. See Pliny, *Natural History,* tr. W. H. S. Jones, second edition (Cambridge, Mass., 1980), vol. 7, p. 141 [book XXV, ix, 8].

4. W. J. Berry and H. E. Poole, *Annals of Printing* (London, 1966), p. 215, s.v. "1826" on incunabula; cf. Lucien Lefebvre and H.-J. Martin, *The Coming of the Book,* tr. David Gerard (London, 1976), p. 248. Printing appeared earlier (ca. 1000

210

A.D.) in China. See Thomas F. Carter and L. C. Goodrich, *The Invention of Printing in China and Its Spread Westward,* second edition (New York, 1955), pp. 245–250.

5. Ivins, *Prints* (n. 1 above), chapters 1, 2, and 8.

6. Ken Baynes and Francis Pugh, *The Art of the Engineer* (London and Woodstock, N.Y., 1981), pp. 226–227. Bruno Latour ("Visualization and Cognition," in *Knowledge and Society* [Greenwich, Conn., 1986), vol. 6, p. 35, n. 16) observes that "the soul of the machine is a pile of paper."

7. Perception of depth in a picture on a flat surface depends upon a cultural background not shared by all people. See Jan B. Deregowski, "Pictorial Perspective and Culture," *Scientific American* 227 (November 1972), pp. 82–88. Marshall H. Segal et al. (*The Influence of Culture on Visual Perception* [Indianapolis, 1966]) argue that the Western "carpentered" world view reinforces depth perception and makes it hard for us to realize that one must learn the cues of depth perception in two-dimensional representations. For details of the dissemination of methods of perspective construction, see Ivins, *Rationalization* (n. 1 above). For a wide-ranging, critical, and insightful book by a "perspector" (maker of perspective drawings), see Lawrence Wright, *Perspective in Perspective* (London, 1983).

8. Samuel Y. Edgerton, Jr., *The Renaissance Rediscovery of Linear Perspective* (New York, 1975), p. xvii. Pages 42–49 elucidate Alberti's method of constructing a grid with a vanishing point, orthogonals, and transversals. See also Wright, *Perspective* (n. 7 above), pp. 64–70.

9. V. P. Zubov, *Leonardo da Vinci,* tr. D. H. Crouse (Cambridge, Mass., 1968), p. 142.

10. The isometric view—a modification of pictorial perspective suitable for objects with a limited depth of field, such as machine elements—was introduced around 1820 by William Farish, a Cambridge don. As compared to a linear perspective view, an isometric view distorts the appearance of an object, but it is much easier to draw. Informal pictorial sketches are often isometric. See P. J. Booker, *A History of Engineering Drawing* (London, 1963), chapter 11.

11. A modern camera records the same image as the camera obscura of the sixteenth century, the difference being that the camera records the image automatically whereas the image in the camera obscura had to be recorded manually. See Mary

Sayer Hammond, The Camera Obscura: A Chapter in the Pre-History of Photography, Ph.D. dissertation, Ohio State University, 1986.

12. Dürer's dozens of carefully dimensioned drawings of male and female bodies (typically more than 20 dimensions for a single body) were intended to improve artists' abilities to draw accurate human figures. Dürer demonstrated how, starting with three orthogonal views, one could construct an accurate delineation of a head in any position.

13. Booker (n. 10 above), p. 44.

14. Booker (n. 10 above), chapter 14.

15. Yves Deforge, *Le Graphisme Technique: Son Histoire et Son Enseignment* (Seyssel, France, 1981), p. 214.

16. See the reproductions of Watt's drawings in H. W. Dickinson and Rhys Jenkins' *James Watt and the Steam Engine* (Oxford, 1927) and in Baynes and Pugh (n. 6 above).

17. For a discussion of "third angle" (American) and "first angle" (European) projection, see Booker (n. 10 above), chapter 14 and p. 179.

18. US Department of Commerce, Patent and Trade Mark Office, *General Information Concerning Patents* (Washington, D.C., 1978), pp. 14–15.

19. *American Standard Graphical Symbols for Heating, Ventilating, and Air Conditioning* (New York, 1949). Current standards may be found in the (current) *1987 Catalog of American National Standards.*

20. Thomas E. French, *A Manual of Engineering Drawing,* sixth edition (New York, 1941); see Thomas E. French, Charles J. Vierck, and Robert J. Foster, *Engineering Drawing and Graphic Technology,* thirteenth edition (New York, 1986), p. 366.

21. Kathryn Henderson, "Flexible Sketches and Inflexible Data Bases: Visual Communication, Conscription Devices, and Boundary Objects in Design Engineering," *Science, Technology, and Human Values* 16, no. 4 (autumn 1991), pp. 448–473, esp. 453, 460–462.

22. Kathryn Henderson, letter to author, December 21, 1990.

23. James M. Edmonson, *From Mécanicien to Ingenieur: Technical Education and the Machine Building Industry in Nineteenth-Century France* (New York, 1987), pp. 306–325.

24. Ibid., p. 309.

25. John Fitchen, *The Construction of Gothic Cathedrals* (Oxford, 1961), pp. xi, 5, 199, 302; L. T. Courtney and Robert Mark, "The Westminster Hall Roof: A Historiographic and Structural Study," *Journal* (Society of Architectural Historians) 46 (December 1987), pp. 374–393; letters to editor taking issue with the article, ibid. 47 (September 1988), pp. 321–324. The article claimed to have "clearly resolved the major questions raised by previous studies" of the hammer-beam roof. A Cambridge engineer objected (p. 322): "I do not believe this to be so. It is all too easy to find what one is looking for and to direct scientific investigations, whether experimental or theoretical, toward predetermined results."

26. Frank D. Prager and Gustina Scaglia, *Brunelleschi: Studies of His Technology and Inventions* (Cambridge, Mass., 1970), pp. 3, 26–32, 42.

27. Bern Dibner, *Moving the Obelisks* (Norwalk, Conn., 1952; Cambridge, Mass., 1970), p. 24.

28. Catherine Brisac, *Le Musée des Plans-Relief, Hotel national des Invalides* (Paris, 1981) is a museum booklet of 100 pp. that describes the origins of the collections. On the status of the 102 extant models in 1987 (76 of which had been "kidnapped" by the Mayor of Lille), see *New York Times*, January 7, 1987, p. 4.

29. Letter of Vauban to Le Peletier de Soucy, October 6, 1695, displayed at Vauban Expo, 1983–84, Musée national des monuments français, Palais de Chaillot, Metro Trocadero. My translation.

30. Howard I. Chapelle, *The Search for Speed Under Sail* (New York, 1967), pp. 150–151, 322–324.

31. Richard G. Hewlett and Francis Duncan, *Nuclear Navy* (Chicago, 1974), p. 173.

32. An experienced worker of the same culture as the designer might well be able to correct obvious mistakes in a copy. However, if the copy is copied, and so on, the original intention of the designer will probably at some point be entirely lost. Samuel Y. Edgerton, Jr., demonstrates how French drawings (Ramelli, 1588) were

completely misunderstood by Chinese copyists in 1627 and 1726. See his "The Renaissance Artist as Quantifier," in *The Perception of Pictures*, ed. M. A. Hazen (New York, 1980). Sources of manuscript drawings: Francesco's originals: (figure 4.23) British Museum, Codex 197 b21, folio 12 recto; (figure 4.25) same ms., folio 30r; (figure 4.27) same ms., folio 42 recto. Anonymous Sienese copyists: (figure 4.24) Biblioteca Nazionale, Firenze, Codex Maglia bechiano, II.III.314 (7646), folio 42 recto; (figures 4.26 and 4.28) same ms., folio 9 recto. Gustina Scaglia kindly located these for me and supplied some of the photographs.

The object being lifted by the copied crane is a column capital, also distorted by the copyist. The top (visible) surface and the bottom surface of the cylinder should appear to be parallel.

Chapter 5

1. Gardner C. Anthony, *An Introduction to the Graphic Language* (New York, 1922), p. iii.

2. Thomas Edison Papers, West Orange, N.J., E1676, Cat. 1172, notebook, p. 24, 29–30 July 1871; reproduced in *The Papers of Thomas A. Edison*, ed. R. Jenkins et al. (Baltimore, 1989), vol. 1, pp. 322–324.

3. Systematic displays of "mechanical movements" were first published in Paris by instructors at the Ecole Polytechnique (see figure 5.6), then imitated and expanded many times before 1871. Two pages of an 1868 book of mechanical movements are shown in figures 5.1 and 5.2.

4. Jacques Besson, *Theatre des instrumens mathematiques et* méchaniques (Lyon, 1578); Jean Errard, *Le premier livre des instruments mathematiques méchaniques* (Nancy, 1584). Eleven editions of Besson's book are listed in Martha Teach Gnudi and E. S. Ferguson, *The Various and Ingenious Machines of Agostino Ramelli* (London and Baltimore, 1976; New York and Aldershot, 1987).

5. Gnudi and Ferguson (n. 4 above).

6. In the bibliography of Gnudi and Ferguson (n. 4 above), the following theatres of machines are listed: Ambroise Bachot, *Le Gouvernail* (Melun, 1598); Vittorio Zonca, *Novo teatro di machine et edificii* (Padua, 1607); Heinrich Zeising, *Theatri machinarum* . . . (Leipzig, 1613–1629); Fausto Veranzio, *Machinae novae* (Florence, 1615); Jacob de Strada, *Künstlicher Abriss*. . . (Frankfurt, 1617–18); George

Böckler, *Theatrum machinarum novum* (Nuremberg, 1661); Gaspard Grollier de Servière, *Recueil d'ouvrages curieux de mathematique et de mécanique* (Lyon, 1719).

7. Jacob Leupold, *Theatrum Machinarum* (10 vols.) (Leipzig, 1724–1739).

8. Vannoccio Biringuccio, *De la pirotechnia* (Venice, 1540; English translation by Cyril Stanley Smith and Martha Teach Gnudi [New York, 1942, 1959, 1990]); Georg Agricola, *De re metallica* (Basel, 1556; English translation by Herbert and Lou Henry Hoover [London, 1912; New York, 1950]).

9. Cyril Stanley Smith, in "Some Important Books in the History of Metallurgy," *Metals Review* 36, no. 9 (1963), pp. 10–14, lists 33 titles published between 1534 and 1901, including those of Lazarus Ercker (1580) and Ciriacus Schreittmann (1578) on smelting and assaying, one by Mathurin Jousse (1627) on iron and steel for lockmaking, one by Jean Boisard (1692) on the minting of coins, and one by Petrus van Musschenbroeck (1729) on testing machines and tensile tests of wire.

10. Joseph Moxon, *Mechanick Exercises or the Doctrine of Handy-Works,* ed. Charles F. Montgomery and Benno Forman (New York, 1970); Joseph Moxon, *Mechanick Exercises on the Whole Art of Printing,* second edition, ed. H. Davis and H. Carter (Oxford and New York, 1962, 1980).

11. Paolo Rossi, *Philosophy, Technology, and the Arts in the Early Modern Era,* tr. Salvator Attanasio (New York, 1970), pp. 5–7.

12. Quoted in Brooke Hindle, *Emulation and Invention* (New York, 1981), p. 13, from Franklin's "Proposals Relating to the Education of Youth" [1749], in *The Papers of Benjamin Franklin,* ed. L. W. Labaree et al. (New Haven, 1961), vol. 3, p. 418.

13. Quoted by Rossi (n. 11 above), pp. 1–2.

14. H. W. Turnbull, ed., *The Correspondence of Isaac Newton, 1661–1675* (Cambridge, 1959), vol. 1, pp. 9–13. For some "useful Hints to Gentlemen that intend to send their sons Abroad to Travel," stressing the value of observing machinery of all kinds, see *Universal Magazine of Knowledge and Pleasure* (London), vol. 2 (1748), p. 247.

15. W. E. Houghton, Jr., "The History of Trades: Its Relation to Seventeenth-Century

Thought," in *Roots of Scientific Thought,* ed. P. Wiener and A. Noland (New York, 1957).

16. Francis Bacon, *The Two Books of the Proficience and Advancement of Learning* [1605] (New York, 1970), book 2, p. 10.

17. J. I. Cope and H. W. Jones, eds., *History of the Royal Society* (St. Louis, 1958), p. 392. Benno Forman, in his introduction to a reprint of Joseph Moxon's *Mechanick Exercises* (n. 10 above), compared Moxon's book to the Royal Society's history of trades when he commented: "As is often the case, the initiative of one man— Joseph Moxon—accomplished what a whole society of well-meaning gentlemen could not."

18. Rossi (n. 11 above), p. 5.

19. Houghton (n. 15 above), p. 368.

20. Houghton (n. 15 above), p. 378: "The quickened interest of the years from 1664 to 1667 had again died down." In 1671, Boyle wrote an essay attempting to revive interest, but (p. 381) "Boyle's design was never achieved."

21. Arthur M. Wilson, *Diderot* (New York, 1972), pp. 242–43; Denis Diderot, ed., *Encyclopédie ou dictionnaire raissoné des sciences, des arts, et des métiers* (Paris, 1751–1780). The original edition consisted of 35 volumes in large folio.

22. *Descriptions des arts et métiers, faites ou approuvées par messieurs de l'Académie Royale des Sciences* (Paris, 1761–1788). The Smithsonian Institution Library copy is bound in 45 volumes.

23. John U. Nef, *The Rise of the British Coal Industry* (London, 1932), vol. 1, p. 242.

24. Walter G. Endrei, "The First Technical Exhibition," *Technology and Culture* 9, no. 2 (1968), pp. 181–183.

25. Académie des Sciences, *Histoire,* 1 (1666–1686) (Paris, 1733), p. 70.

26. Académie des Sciences, *Machines et inventions approuvée par* l'Académie Royale des Sciences (Paris, 1735–1777); *Roger Hahn, The Anatomy of a Scientific Institution: the Paris Academy of Sciences, 1666–1803* (Berkeley, 1971), p. 23.

27. Derek Hudson and Kenneth W. Luckhurst, *The Royal Society of Arts 1754–1954* (London, 1954), pp. 6–20, 112, 113; William Bailey, *The Advancement of Arts, Manufactures, and Commerce, or, Descriptions of the Useful Machines and Models Contained in the Repository of the Society for the Encouragement of Arts, Manufactures, and Commerce* (London, 1776). See also Kenneth W. Luckhurst, *The Story of Exhibitions* (London and New York, 1951), pp. 63–69. Greville and Dorothy Bathe's book *Oliver Evans* (Philadelphia, 1935) reprints a 1793 letter sent to Evans from London by a Philadelphia clock and watch maker, Robert Leslie: "I have got acquainted with several of the members of the Society and soon expect to have admittance to see their collection of machines, and models, which is said to be the greatest in the world." (p. 41)

28. Svante Lindqvist, *Technology on Trial: The Introduction of Steam Power Technology into Sweden, 1715–1735* (Uppsala, 1984), chapter 2, "Wooden Images of Technology."

29. Notebooks of Carl J. Cronstedt, part 3 (1729), in Tekniska Museet, Stockholm. Early in the nineteenth century, Robert Fulton of steamboat fame also expressed the idea of a mechanical alphabet. Brooke Hindle (n. 12 above, p. 135) quotes him: "The mechanic should sit down among levers, screws, wedges, wheels, etc. like a poet among the letters of the alphabet, considering them as the exhibition of his thoughts, in which a new arrangement transmits a new Idea to the world."

30. R. Tresse, "Le Conservatoire des Arts et Métiers et la Societé d'Encouragement pour l'Industrie nationale au debut du XIXe siècle," *Revue d'histoire des sciences et de leurs applications* 5, no. 3 (1952), pp. 246–264.

31. Eugene S. Ferguson and Christopher Baer, *Little Machines: Patent Models in the Nineteenth Century* (Wilmington, Del., 1979), pp. 6–11. In 1836 a fire destroyed all the records and the models. Another fire, in 1877, destroyed about one-third of the 225,000 models that had been accumulated since 1836.

32. A. E. Musson and Eric Robinson, *Science and Technology in the Industrial Revolution* (Manchester, 1969), pp. 37–45, 103, 105, 111, 180, and 181; see also "Scientific Lectures" in the index.

33. John T. Desaguliers, *A Course of Experimental Philosophy* (London, 1734–1744); Benjamin Martin, *Philosophia Britannica* (London, 1747); James Ferguson, *Lectures on Select Subjects* (London, 1764). The pattern of the English lectures originated in France. Probably the most direct influence was Jacques Ozanam's

five-volume *Cours de mathematiques* (Paris, 1693), volumes 4 and 5 of which were translated into English by Desaguliers (London 1712). Biographies of Ozanam and Desaguliers appear in the *Dictionary of Scientific Biography* (New York, 1970–1980).

34. Hindle (n. 12 above), pp. 30, 34, 60.

35. I. Bernard Cohen, *Some Early Tools of American Science* (Cambridge, Mass., 1950), pp. 158, 162, 163; David P. Wheatland, *The Apparatus of Science at Harvard, 1765–1800* (Cambridge, Mass., 1968), pp. 86–87, 98–99, 102–103.

    Massachusetts Magazine (January 1790, p. 63) announced the arrival of a "Philosophical Apparatus" from London, in good order, for Yale College, adding that "the University is thus furnished with a complete set of instruments and machines, for exhibiting a whole course of experiments in Natural Philosophy and Astronomy." The apparatus was purchased with liberal donations from several gentlemen, Rev. Dr. Lockwood being the principal donor.

36. Franz Reuleaux, *The Kinematics of Machinery,* tr. A. B. W. Kennedy (London, 1875; New York, 1963), p. 14. Kennedy (*Engineering* 22 [1876], 239) introduced the models to England.

37. Maurice d'Ocagne, *Le Calcul Simplifié: Graphical and Mechanical Methods for Simplifying Calculation,* tr. J. Howlett and M. R. Williams (Cambridge, Mass., 1986), pp. 155–158.

38. d'Ocagne (n. 37 above), pp. 113–115, 118. S.v. Nicholas Oresme in *Dictionary of Scientific Biography* (n. 33 above), vol. 10, pp. 223–230, esp. p. 226.

39. Indicator diagrams were used to analyze the operation of reciprocating engines. They were drawn by a small instrument called an "indicator," comprising a cylinder, a piston, and an oscillating drum carrying a paper "indicator card." Pressure was measured vertically, volume horizontally. Figure 5.23 shows an "ideal" indicator card. Pathological deviations from the ideal (figure 5.24) were diagnosed in the textbook in which they appeared: E. V. Lallier, *Elementary Steam Engineering* (New York, 1926), pp. 123, 129. The indicator diagram is a useful but wholly empirical curve whose shape presents a formidable problem in determining mathematically the mean pressure during the cycles, from which a power output can be derived. Via graphical procedures, the problem is solved by a planimeter, which integrates the area as a point on the instrument's arm is traced around the diagram. Since the ordinate represents pressure and the abscissa volume, the diagram repre-

sents the integral of *pdv,* or the work done during a cycle of the engine. Knowing the cycles per minute, one can readily determine the horsepower.

40. Karl Culmann, *Graphische Statik* (Zurich, 1866). A. J. Dubois (*The New Method of Graphical Statics* [New York, 1875]) explained Culmann's method in English. A series of articles had appeared earlier in *Van Nostrand's Engineering Magazine.*

41. Elwin C. Robison, letter to author, January 9, 1990.

42. d'Ocagne (n. 37 above). See the editors' "New Introduction for This Edition," pp. ix–xiv.

43. A mid-twentieth-century handbook of solved equations is *Falk's Graphical Solutions to 100,000 Practical Problems,* by Karl H. Falk (Columbia, Conn., 1946).

Chapter 6

1. Quoted in John H. Weiss, *The Making of Technological Man: The Social Origins of French Engineering Education* (Cambridge, Mass., 1982), p. 165. Olivier concluded with the thought that they "should have changed the name of the school from polytechnique to monotechnique."

2. Letter to the editor, *Mechanical Engineering* 113 (March 1991), p. 6.

3. After literary and mathematical schooling, Brunelleschi enrolled in the guild of goldsmiths, bronze workers, and metal workers. He was apprenticed in 1392 at age 15 and qualified as a master in 1398. See Eugenio Battisti, *Brunelleschi: The Complete Work* (London, 1981), p. 22.

4. Agostino Ramelli, *Various and Ingenious Machines* [1588], tr. M. Teach Gnudi (London and Baltimore, 1976; New York and Aldershot, 1987), pp. 46–53.

5. Lynn White, Jr., "The Flavor of Early Renaissance Technology," in *Developments in the Early Renaissance,* ed. B. A. Levy (Albany, N.Y., 1972), quotation on p. 47.

6. Walter E. Houghton, "The History of Trades: Its Relation to Seventeenth-Century Thought," in *Roots of Scientific Thought,* ed. P. Wiener and A. Noland (New York, 1957), esp. pp. 355–359.

7. Francis Bacon, *The Great Instauration and the New Atlantis* (Arlington Heights, Ill., 1980), pp. 55–57, 79–80. The king of New Atlantis was named Salomona.

8. The myth was restated by Vannevar Bush on pp. 52–53 of his *Endless Horizons* (Washington, D.C., 1946).

9. The components of Morse's first telegraphic instrument (preserved in the Smithsonian Institution) were mounted on a painter's canvas stretcher. As the historian Brooke Hindle has observed (*Emulation and Invention* [New York, 1981], pp. 120–121), the physical form of the telegraph might have been "grossly different" otherwise. According to Hindle, Morse used the stretcher "because it was there."

10. Ibid.

11. Brooke Hindle, "The Underside of the Learned Society in New York, 1754–1854," in *The Pursuit of Knowledge in the Early American Republic*, ed. A. Oleson and S. C. Brown (Baltimore, 1976), esp. pp. 86–92.

12. Donald Fleming, *John William Draper and the Religion of Science* (Philadelphia, 1950), pp.1, 139–142.

13. Charles Rosenberg, "Science and American Social Thought," in *Science and Society in the United States*, ed. D. T. Van Tassel and M. G. Hall (Homewood, Ill., 1966), esp. p. 153.

14. Marcel C. La Follette, *Making Science Our Own: Public Images of Science, 1910–1955* (Chicago, 1990).

15. *The Memoirs of Herbert Hoover: The Cabinet and the Presidency 1920–1933* (New York, 1952), pp. 112–124.

16. Vannevar Bush, *Pieces of the Action* (New York, 1970), pp. 53–54.

17. Ibid., pp. 53–54. Bush thought the contempt for engineers was more marked among British officers (with whom he had much to do) than among US officers.

18. Vannevar Bush, *Endless Horizons* (n. 8 above), pp. 52–53. If Bush believed this paragraph, he was captive to a myth he had grown up with during the Progressive era.

A Department of Defense study (1958–1966) called "Hindsight," designed to justify expenditures for "pure" research, revealed that of 700 key contributions to weapons systems, 91% were technological, 8.7% were "applied science," and only 0.3% resulted from "pure" undirected research. Edwin Layton discusses the implications of this in his article "Mirror-Image Twins: The Communities of Science and Technology in 19th-Century America," *Technology and Culture* 12, no. 4 (October 1971), pp. 562–580.

19. Bush (n. 18 above), p. 53.

20. Ibid., p. 74.

21. Paul Forman, "Behind Quantum Electronics: National Security as Basis for Physical Research in the United States, 1949–1960," *Historical Studies in the Physical and Biological Sciences* 18, no. 1 (1987), pp. 149–229, esp. 156. See also Herbert F. York and G. Allen Gelb, "Military Research and Development: A Postwar History," in *Science, Technology, and National Policy,* ed. T. J. Kuehn and A. L. Porter (Ithaca, 1981).

22. Harvey M. Sapolsky, "Academic Science and the Military: The Years since the Second World War," in *The Sciences in American Context: New Perspectives,* ed. N. Reingold (Washington, D.C., 1979), esp. p. 379.

23. Forman (n. 21 above, p. 229) shows in detail how in the postwar years physicists, "though they . . . maintained the illusion of autonomy with pertinacity, . . . lost control of their discipline." Members of engineering faculties were and are expected not only to do sponsored research but also to find the money required to support it. In a peremptory article ("Funded Research: Getting Started," *Journal of Engineering Education* 80 (January–February 1990), pp. 27–31), Marshall H. Kaplan advises the faculty researcher to find a "funding 'patron saint,' [a government] agency employee [who] not only believes that [the supplicant's] research will help fulfill his or her objectives, but also has the funding to support the effort."

24. "Interim Report of the Committee on Evaluation of Engineering Education," *Journal of Engineering Education* 45 (September 1954), pp. 40–66, esp. 43.

25. Ibid., p. 40. See also "Selected Reports by Institutional Committees on Evaluation of Engineering Education," *Journal of Engineering Education* 44 (December 1953), pp. 257–272.

26. Quoted from the separately published report *Evaluation of Engineering Education, 1952–1955* (Urbana, Ill., 1955) in Eric A. Walker and Benjamin Nead, "The Goals Study," *Engineering Education* 57 (September 1966), pp. 13–19; quotation on p. 16.

27. [Final] "Report of the Committee on Evaluation of Engineering Education," *Journal of Engineering Education* 46 (September 1955), pp. 25–60, esp. 42.

28. See Robert H. Thurston, "Aim and Scope of the [Mechanical] Engineering College," *Engineering Magazine* 10 (1895–96), pp. 418–422, esp. p. 420. Thurston outlines the "manual training" received by a student at Cornell and justifies it thus: "All this is done, not to make him a good mechanic in the usual sense, or to teach him a trade, but to enable him to design intelligently, to criticize good and bad work, to indicate the best methods of doing work directed by him, and to guide the workman intelligently."

29. Final Report (n. 27 above), p. 37.

30. Final Report (n. 27 above), "Summary," p. 25. The problem of finding qualified faculty members to teach such "integrated courses" remains unsolved. The enthusiasm of young Ph.D. instructors cannot replace the seasoning in engineering practice and design that effective instructors in engineering design must have.

31. Paul F. Chenea, "Teaching Design for the Rest of the Twentieth Century," *Journal of Engineering Education* 51 (March 1961), pp. 566–570, esp. p. 566.

32. B. R. Teare, Jr., "Design," in *Britannica Review of Developments in Engineering Education*, ed. N. Hall (Chicago, 1970), esp. p. 138.

33. "Report on Engineering Design" [MIT Committee on Engineering Design], *Journal of Engineering Education* 51 (April 1961), pp. 645–660; quotations from p. 649.

34. Ibid., pp. 645–646.

35. Ibid., p. 651.

36. Ibid., p. 651.

37. Ibid., pp. 647–648.

222

38. Ibid., pp. 647–649.

39. Ibid., p. 650.

40. Ibid., p. 652. This was prefaced "It seems trite but perhaps necessary to say that. . . ."

41. Ibid., p. 652.

42. A 160-page report of the survey was published in 1980 by the engineering department of Cambridge University.

43. Ibid., pp. 19, 20.

44. Ibid., p. 136.

45. Ibid., p. 45.

46. Ibid., pp. 22–44.

47. Engineering technology evolved from "engineering associate," a two-year certificate program established in many engineering schools after 1945. See Lawrence J. Wolf, "The Emerging Identity of Engineering Technology," *Engineering Education* 77 (April–May 1987), pp. 725–734.

48. In 1988–89, the schools granting more than 200 B.S.E.T. degrees each were Southern College of Technology (Georgia), Devry Institute (Chicago), Purdue University, Northeastern University, Wentworth Institute of Technology (Boston), Rochester Institute of Technology, Devry Institute (Columbus, Ohio), Oregon Institute of Technology (Klamath Falls), Pennsylvania State University, Texas A&M University, and Old Dominion University. The total number of B.S.E.T. degrees, conferred by 137 schools in the United States, was 11,289. See Richard A. Ellis, "Engineering and Engineering Technology Degrees, 1988," *Engineering Education* 80 (April 1990), pp. 413–422.

49. For 1989, see Ellis (n. 48 above); for engineering degrees, 1976–1988, see Ellis, "Engineering and Engineering Technology Degrees, 1988," *Engineering Education* 79 (May–June 1989), p. 511. For a table of engineering degrees conferred in the years 1949–1971, see *Engineering Education* 62 (April 1972), p. 800.

50. Stanley W. Anderson, "The Technology Graduate Today," in American Society for Engineering Education, *Proceedings 1979, College-Industry Education Conference* (Tampa, 1979). Of the 57 technologists, 17 were in production (i.e., generating stations—and distribution?), 14 in "operational analysis," 7 in marketing, 5 in station construction, and the rest in quality assurance, purchasing, and various engineering functions. None was in a design division.

In the same proceedings (p. 240), the Dean of Engineering at the New Jersey Institute of Technology wrote: "Up until [1972] I did not know what Engineering Technology was. I did know that there was a need for technically trained people in the applications of engineering principles and that, in general, the engineering education was not filling the need. It is rewarding to know that our product *is* filling a need, *is* accepted by industry . . . and the graduates *are* performing well in their jobs."

51. Richard C. Mallonee II, An Historical Analysis of Major Issues and Problems in the Development of the Baccalaureate Degree in Engineering Technology, Ph.D. dissertation, University of Washington, 1979. On licensing, see pp. 10–11; on attempts by professional societies to restrict employment opportunities of ET graduates, see passim. On ABET restrictions, see Ron Williams, "A Question of Ethics," *Engineering Education* 72 (April 1972), p. 694.

52. Wolf (n. 47 above). Wolf's article, which asserts that "engineering does not lose when engineering technology wins," catalogues several standard objections to the BSET. He points out, however, that the National Center for Education Statistics (publishing in the *Chronicle of Higher Education*) counted 17,022 BSET degrees in 1982–83 while the ASEE counted only 10,200.

53. "Report on Engineering Design" (n. 33 above), p. 649.

54. This conclusion is supported in a brief 1989 article by Eric A. Walker, president emeritus of Pennsylvania State University, former president of National Academy of Engineering, and chairman of the 1962–1965 ASEE evaluation committee that produced a report entitled "Goals of Engineering Education" (*Engineering Education* 58, no. 5 [January 1968]). See Eric A. Walker, "Our Engineering Schools Must Share the Blame for Declining Productivity," *Chronicle of Higher Education* 34 (December 2, 1987), p. 52. Walker asserts that although "nearly 80,000 new engineers were graduated last year," "only a minority of them left college knowing how to design and manage production—in other words, how to 'do' engineering."

224

Chapter 7

1. Robin D. Higham, *The British Rigid Airship, 1908–1931* (London, 1961), quoted in Arthur M. Squires, *The Tender Ship* (Boston, 1986), p. 7.

2. *New York Times,* December 6, 1990.

3. Jan Adkins, *Moving Heavy Things* (Boston, 1980), p. 9.

4. S. S. Baxter and M. Barkofsky, "Construction of the Walnut Lane Bridge," in Proceedings of the First United States Conference on Prestressed Concrete (MIT, August 14–16, 1951), pp. 47–49.

5. *Civil Engineering* 47 (October 1977), p. 119.

6. *New York Times,* January 17, 1990.

7. Hubert L. and Stuart E. Dreyfus (*Mind Over Machine* [New York (paper), 1988], p. 69) mention "the basic unsolved problem of AI [artificial intelligence]." See their index, s.v. common sense; the 20 citations are worth running down. Douglas Hofstadter is quoted on p. 67: "Yet there is no program that has common sense; no program that learns things it has not been explicitly taught how to learn; no program that can recover gracefully from its own errors."

8. James Gleick, *Chaos: Making a New Science* (New York, 1987), p. 210.

9. Ibid., pp. 278–279.

10. Ibid., pp. 15–23.

11. Alan Colquhoun, "Typology and the Design Method," *Perspecta: The Yale Architectural Journal* 12 (1969), pp. 71–74.

12. Robert W. Mann, "Engineering Design," in *McGraw-Hill Encyclopedia of Science and Technology* (New York, 1989), vol. 6, p. 359.

13. *Engineering News* 17 (April 9, 1887), pp. 229, 237–238.

14. A recent book based upon articles in *Engineering News-Record* is Steven S. Ross'

Construction Disasters: Design Failures, Causes, and Prevention (New York, 1984).

15. *Engineering News* 58 (September 5, 1907), pp. 256–257.

16. Ibid.

17. *Scientific American* 97 (October 12, 1907), pp. 257–258.

18. Steven S. Ross, *Construction Disasters: Design Failures, Causes, and Prevention* (New York, 1984), pp. 377–388.

19. *Science* 204 (April 7, 1989), p. 29.

20. Henry Petroski, *To Engineer Is Human: The Role of Failure in Successful Design* (New York, 1985).

21. Ibid., pp. 198–200. For additional details, see Ross (n. 14 above), pp. 303–322.

22. The space frame of the Kansas City roof did not fail, but steel bolts connecting it to the roof deck suspended beneath it failed and the roof collapsed. See Ross (n. 14 above), pp. 322–344.

23. *Engineering News-Record* 205 (September 18, 1980), p. 39; Ross (n. 14 above), p. 322.

24. Petroski (n. 20 above), pp. 201–202.

25. *Engineering News-Record* 202 (April 5, 1979), pp. 10–15, esp. p. 13. See also Charles Perrow, "Normal Accident at Three Mile Island," *Society* 18, no. 5 (July–August 1981), pp. 17–26, esp. pp. 21–22. See also Perrow's *Normal Accidents: Living With High-Risk Technologies* (New York, 1984). "An Analysis of Three Mile Island" (staff-written) appears in *IEEE Spectrum* 16 (November 1979), pp. 32–34. Although all instruments display inferential information (e.g., in a thermometer, a length of mercury column to indicate temperature), a fail-safe indicator is nearly always tied directly to the condition being reported.

26. *New York Times,* January 6, 10, and 11 and February 8, 1990.

226

27. *New York Times,* April 12, 1990.

28. *New York Times,* May 30, 1989.

29. *Wall Street Journal,* May 7, 1990.

30. Ibid.

31. For another recent US product failure, see "How US Robots Lost the Market to Japanese in Factory Automation," *Wall Street Journal,* November 6, 1990. US makers clung to hydraulic robots long after the superiority of electrical robots became evident.

32. National Aeronautics and Space Administration, *The Space Telescope* (Washington, D.C., 1976) [call no. NAS 1.21:392], p. 51: "This Earth-orbiting observatory will open a new era of astronomy because it can see 7 times farther and 350 times as much volume as the best ground-based telescope. It also has 10 times better resolution and 10 times the frequency spectrum of ground-based systems."

33. *Science* 249 (July 6, 1980), p. 25; *New Scientist* 127 (September 27, 1990), p. 30.

34. *New York Times,* May 1, May 10, June 28, and June 29, 1990.

35. *New York Times,* June 15, 1990.

36. *Sky and Telescope* 82 (1991), pp. 239, 246, 350, 581.

37. *New York Times,* March 19, 1990.

38. *New York Times,* March 19, 28, and 29 and June 10, 1990.

39. Richard P. Feynman, *"What Do You Care What Other People Think?"* (New York, 1988), pp. 226–232.

40. Ibid., p. 214.

41. Alfred Pugsley, *The Safety of Structures* (London, 1966), pp. 141–144, 150.

42. Ibid., pp. 144–147, 150.

43. Finch published "Wind Failures of Suspension Bridges or Evolution and Decline of the Stiffening Truss" in *Engineering News-Record* (126 [March 13, 1941], pp. 402–407). Two weeks later (ibid. [March 27], p. 43), he wrote: ". . . attention has been called to the fact that the casual reader may infer . . . that the modern bridge engineer, in view of the earlier failures of bridges, was remiss. . . . The author . . . did not suggest or intend the reader to imply, that the modern engineer should have known the details of the earlier disasters. . . ." Such are the pressures when one is publishing criticisms of engineering works.

44. David P. Billington, "History and Esthetics in Suspension Bridges," *Journal of the Structural Division* [ASCE] 103 (August 1977), pp. 1665–1672; discussion in ibid. 104 (1978), pp. 246–249, 378–380, 619, 732–733, 1027–1035, 1174–1176; closure in 105 (March 1979), pp. 671–687. Billington's paper was criticized angrily by five discussants.

45. David P. Billington, *The Tower and the Bridge* (New York, 1983), p. 137. The trauma of the moment was reflected in the report in *Engineering News-Record*. In the official report of the installation of the stiffening trusses, all the drama of the experience was expunged; a reader might believe that the rebuilding of the bridge was a routine affair. See *The Golden Gate Bridge. Report of the Chief Engineer* (1937) and *Supplement to the Final Report* by Clifford E. Paine (bound together) (San Francisco, 1970). An informative book on planning and building the bridge is John Van Der Zee's *Gate: The True Story of the Golden Gate Bridge* (New York, 1986).

46. Billington (n. 45 above), p. 137. Billington pointed out that Ammann "worked in an age captivated by the idea of mathematical science as a prerequisite to engineering practice. The creation of new designs based upon earlier experience seemed to become less useful once people began to believe that new works would come from new research, that scientific discoveries would produce technological 'breakthroughs.'" We should remember that we are still living in that age.

47. Herbert Rothman's "Discussion" of Billington's 1977 article (note 43 above) appears in *Journal of the Structural Division* 104 (January 1978), pp. 246–249.

48. *Scientific American* 259 (September 1988), pp. 14, 18.

49. See "A Case of Human Error," *Newsweek,* August 15, 1990, pp. 18–20. An editorial in the *New York Times* of August 4, 1988, reminded readers that the

Navy's verdict in the attack on the *Stark* (May 1987) was "crew, guilty of misjudg-ment; equipment blameless," and that in the *Vincennes* affair there were said to have been "no problems with the Aegis." The verdict of the *Times* was "any equipment that puts its trained crew so wholly in the wrong may need thorough overhaul, whatever the inconvenience, cost or loss of face." For a thoughtful review of the limits of "systems" thinking in man-machine systems, see Franklin V. Taylor, "Four Basic Ideas in Engineering Psychology," *American Psychologist* 15, no. 10 (October 1960), pp. 643–649.

Notes to the Figures

f1. Fay L. Faurote, "Equipment Makes Possible the Ford Model A Aluminum Piston and Extruded Piston Pin," *American Machinist* 68 (May 10, 1928), pp. 761–762.

f2. Wing warping is described in F. E. C. Culick, "The Origins of the First Powered, Man-Carrying Airplane," *Scientific American* 241 (July 1979), pp. 86–100. See also Orville Wright, *How We Invented the Airplane* (New York, 1988). Figure 1.5 is from the latter (p. 14).

f3. Frontispiece of Newcomen Society's *Transactions* 4 (1923–24).

f4. Otto von Guericke, *Experimenta Nova (ut vocantur) Magdeburgica de Vacuo Spatio* (Amsterdam, 1672).

f5. Académie des Sciences (1666–1699) 9, p. 434; see also pp. 247–251. Also published in Venturus Mandey and James Moxon, *Mechanick-powers: or, The Mistery of nature and art unvail'd* (London, 1696).

f6. Leonardo da Vinci, Ms. B, f. 54v.

f7. Leonardo da Vinci, Ms. F, 16v.; see Ivor B. Hart, *The World of Leonardo da Vinci* (New York, 1961), pp. 299–300.

f8. *Oeuvres complètes de Christiaan Huygens* (La Haye, 1897), vol. 7, p. 357 (September 22, 1673).

f9. H. W. Dickinson, *A Short History of the Steam Engine* (Cambridge, 1938), pp. 8–10. Papin's steam cylinder was published in *Acta Eruditorum*, 1690, pp. 410–414.

f10. Snapshot taken in Oklahoma, 1973.

f11. "Goodyear Proposes Expandable Structures as Space Stations," *Missiles and Rockets* 8 (May 29, 1961), p. 24.

f12. John Van der Zee, *Gate: The True Story of the Design and Construction of the Golden Gate Bridge* (New York, 1986); Stephen Cassady, *Spanning the Gate* (Mill

Valley, Calif., 1979). The illustration is from *Scientific American* 131 (October 1924), p. 258.

f13. *US Patent Office Gazette,* August 17, 1915. The negative has been reversed to emphasize similarities of shape.

f14. Golden Gate Bridge and Highway District, *The Golden Gate Bridge: Report of the Chief Engineer,* September, 1937, bound with *Supplement* dated December 1, 1970, containing "supplementary information on modifications to the structure since opening day." This drawing is from plate I.

f15. Source: David P. Billington, *Robert Maillart's Bridges, the Art of Engineering* (Princeton, 1979), p. 20.

f16. Ibid., p. 36.

f17. The cylindrical phonograph is from *Scientific American* 38 (March 30, 1878), p. 193. The plate machine is from a sketch in the Edison Papers at West Orange, N.J. (unbound laboratory notebook, vol. 17; reproduced in part I of Edison Papers Microfilm Edition, reel 4, frame 897).

f18. This photograph, in the Smithsonian Institution, was called to my attention by Robert M. Vogel.

f19. *Mechanical Engineering* 83 (December 1961), p. 78.

f20. Nigel Cross, *Engineering Design Methods* (New York, 1989), p. 21.

f21. This, the first engraving of Evans' mill, appeared in the Philadelphia journal *Universal Asylum and Columbian Magazine,* January 1791.

f22. More kinematic details of the parallel motion may be found in my "Kinematics of Mechanisms from the Time of Watt," US National Museum *Bulletin* 228 (Washington, 1963), pp. 185–231, esp. 193–198. Years after developing the parallel motion, Watt told his son: "Though I am not over anxious after fame, yet I am more proud of the parallel motion than of any other mechanical invention I have ever made." (ibid., pp. 197–198). This drawing is from John Farey, *Treatise on the Steam Engine* (London, 1827), p. 444, pl. XI.

f23. James R. Hansen, *Engineer in Charge: A History of the Langley Aeronautical*

Laboratories, 1917–1958 (Washington, 1987), pp. 98–99; "The Characteristics of 78 Related Airfoil Sections from Tests in the Variable-Density Wind Tunnel," in NACA's *19th Annual Report—1932* (Washington, D.C., 1934), pp. 299–351.

f24. Richard T. Whitcomb, "A Study of Zero-Lift Drag-Rise Characteristics of Wing-Body Combinations Near the Speed of Sound," in NACA's *42nd Annual Report—1956*, pp. 519–540.

f25. Hero of Alexandria, *The Pneumatics, A facsimile of the 1851 Woodcroft Edition, Introduced by Marie Boas Hall* (London, 1971), drawing of "The Fire-Engine," p. 44.

f26. John Peter Oleson, *Greek and Roman Mechanical Water-Lifting Devices: The History of a Technology* (Toronto, 1984). The Valverde del Camino pump is described on pp. 268–269 and illustrated in figs. 141–144.

f27. Measured drawing, fig. 142, from J. M. Luzon, "Los sistemas de desagüe en minas romanas del suroeste peninsular," *Archivo Español de Arquelogia* 41 (1968), pp. 101–120.

f28. Brunelleschi's crane is in *Zibaldone* (Bibl. Nationale, Florence, Ms. BR 228, f. 105r).

f29. Leonardo's copy is in his *Codice Atlantico* (Florence and New York, 1973–1975), fol. 808 recto.

f30. Christopher Duffy, *Siege Warfare: The Fortress in the Early Modern World 1494–1660* (London, 1979), p. 33. The drawing is from the US National Archives. See also Richard Welsh, "The Star Fort, 1814," *Maryland Historical Magazine* 54, no. 3 (September 1959), pp. 296–309.

f31. Denis Diderot, *Encyclopédie* (Paris, 1751–1780), Planches v. 1, Art militaire, plate XVII.

f32. *Sketchbook of Wilars de Honecourt*, ed. R. Willis (London, 1859), pl. 43; *The Sketchbook of Villard de Honnecourt*, ed. T. Bowie (Bloomington, Indiana, 1959).

f33. Jacques Besson, *Theatre des instrumens mathematiques et mechaniques* (Lyon, 1578), pl. 14.

f34. Albrecht Dürer, *The Painter's Manual*, tr. W. L. Strauss (New York, 1979), p. 390.

f35. Ibid., p. 434.

f36. Gustina Scaglia, Frank D. Prager, and Ulrich Montag, *Mariano Taccola, De Ingeneis* (Wiesbaden, 1984), vol. 2.

f37. Georg Agricola, *De Re Metallica* [1556], ed. H. and L. H. Hoover (New York, 1950), p. 180.

f38. Leonardo da Vinci, *Il Codice Atlantico* (Florence and New York, 1973–1975), vol. 9, p. 808.

f39. *Motor's Auto Repair Manual* (New York, 1956), p. 746.

f40. Albrecht Dürer, *Hierin sind begriffen vierbücher* [1528] (Bamberg, 1969).

f41. American National Standards Institute, *Graphic Symbols for Heating, Ventilating, and Air Conditioning* Y32.2.4–1949 (R 1984).

f42. Lincoln Electric Company, *How to Read Shop Drawings* (Cleveland, 1961), pp. 149, 150.

f43. Leonardo da Vinci, *The Madrid Codices* (New York, 1974), vol. 1, f. 10v. For a study of Leonardo's thinking while sketching, see Bert S. Hall and Ian Bates, "Leonardo, the Chiaravalle Clock and Epicyclic Gearing: A Reply to Antonio Simoni," *Antiquarian Horology* 9 (September 1976), pp. 910–917.

f44. Smithsonian Institution, National Museum of American History, Erasmus D. Leavitt Papers. Leavitt's press copies were filed for record and reference.

f45. For an interesting narrative of the moving of the Vatican obelisk see Bern Dibner, *Moving the Obelisks* (Norwalk, Conn., 1950; Cambridge, Mass., 1970).

f46. Edison Papers, West Orange, N.J., File E1676, Cat. 1172, July 1871–February 1872, pp. 10–11; in *The Papers of Thomas A. Edison*, ed. R. V. Jenkins (Baltimore, 1989–), vol. 1, pp. 322–324, 343.

f47. Henry T. Brown, *Five Hundred and Seven Mechanical Movements* (New York, 1868; Bronxville, N.Y., 1981), pp. 24, 60.

Notes to Figures

f48. Martha Teach Gnudi and Eugene S. Ferguson, *The Various and Ingenious Machines of Agostino Ramelli* [1588] (New York and Aldershot, 1987), pl. 38; Jacob Leupold, *Theatrum hydraulicarum* (Leipzig, 1725), vol. 1, fig. XLIX, p. 132; John P. Rollins, ed., *Compressed Air and Gas Handbook,* fifth edition (Englewood Cliffs, N.J., 1989), p. 140.

f49. Jean N. P. Hachette, *Traité elémentaire des machines* (Paris, 1811), pl. 1. The panel of kinematic models is one of ten in the Newark (N.J.) Museum.

f50. Mariano Taccola, *De Machinis* [1449], ed. G. Scaglia (Wiesbaden, 1971), vol. 2, p. 196.

f51. Ladislao Reti, "Francesco di Giorgio Martini's Treatise on Engineering and Its Plagiarists," *Technology and Culture* 4, no. 3 (Summer 1963), pp. 287–298.

f52. Bernard F. de Belidor, *Architecture Hydraulique* (Paris, 1739), vol. 2.

f53. E. F. and N. Spon, *Dictionary of Engineering* (London, 1873), fig. 5815.

f54. Will M. Clark, *A Manual of Mechanical Movements* (Garden City, N.Y., 1943).

f55. Denis Diderot, *Encyclopédie* (Paris, 1751–1780), Planches v. 4, s.v. "Epinglier."

f56. "The Physical Laboratory of the Académie des Sciences of Paris, 1711," *Illustrated London News,* May 21, 1938.

f57. Photographs courtesy of Tekniska Museet, Stockholm, and Svante Lindqvist. See chapter 2 of Lindqvist's *Technology on Trial: The Introduction of Steam Power Technology into Sweden, 1715–1736* (Uppsala, 1984).

f58. Harvard University Collections. See Bernard I. Cohen, *Some Early Tools of American Science* (Cambridge, Mass., 1950), pp. 158, 163; David P. Wheatland, *The Apparatus of Science at Harvard* (Cambridge, Mass., 1968), pp. 86, 103.

f59. *Scientific American* 53 (October 17, 1885), p. 239.

f60. Franz Reuleaux, *The Kinematics of Machinery,* tr. A. B. W. Kennedy (London, 1875).

f61. E. V. Lallier, *Elementary Steam Engineering* (New York, 1926), p. 123.

234

f62. Ibid., p. 126.

f63. Thomas E. French and Charles J. Vierck, *Graphic Science* (New York, 1958), p. 647.

f64. Phoenix Bridge Company Papers, Hagley Museum and Library.

f65. Phoenix Bridge Company Papers, Hagley Museum and Library. I am indebted to Chris Baer for locating this photograph.

f66. *Engineering News* 58 (September 5, 1907), p. 258.

f67. Phoenix Bridge Company Papers, Hagley Museum and Library.

f68. *Scientific American* 97 (October 12, 1907), pp. 257–258.

f69. After report prepared by Lev Zetlin Associates.

f70. AP/Wide World Photos.

f71. Jacob Feld summed up the visual message when he wrote in *Construction Failures* (New York, 1968) that "the photographic record of the torsional oscillations made by Professor F. B. Farquharson did more to prove the necessity for aerodynamic investigation of structures than all the theoretical reports."

Index

236

238